IMAGES
of Aviation

JOHNSON
SPACE CENTER
THE FIRST 50 YEARS

Employees gather in the formation of "NASA" in Rocket Park, located near the entrance of the Lyndon B. Johnson Space Center, Houston, Texas. (Courtesy of NASA Johnson Space Center.)

On the Cover: The Apollo Extravehicular Mobility Unit (EMU) provided astronauts with life support, communications, and environmental protection, allowing them to work on the lunar surface in relative comfort despite the harsh conditions. Qualification tests on the EMU, conducted in the man-rated Chamber B of the Space Environment Simulation Laboratory in Building 32, simulated the thermal-vacuum effects anticipated during a lunar stay. The astronauts practiced physically moving and working as well as deploying scheduled lunar experiments. Crew safety and the success of the manned space program depended on information that resulted from these tests. This photograph was captured during testing in 1969. (Courtesy of NASA Johnson Space Center.)

IMAGES
of Aviation

JOHNSON SPACE CENTER
THE FIRST 50 YEARS

Edited by Laura Bruns and Mike Litchfield

ARCADIA
PUBLISHING

Work of the United States Government
Source: The National Aeronautics and Space Administration (NASA) or as otherwise noted.
ISBN 978-0-7385-9510-8

Published by Arcadia Publishing
Charleston, South Carolina

Printed in the United States of America

Library of Congress Control Number: 2012950057

The materials herein identified as "Source: NASA", "Courtesy of NASA Johnson Space Center", or a similar acknowledgement are a work of the U.S. Government and are not subject to copyright protection in the United States.

For all general information, please contact Arcadia Publishing:
Telephone 843-853-2070
Fax 843-853-0044
E-mail sales@arcadiapublishing.com
For customer service and orders:
Toll-Free 1-888-313-2665

Visit us on the Internet at www.arcadiapublishing.com

To the thousands of individuals who provided the nation an avenue to space and set a path for future generations to carry the legacy forward.

Contents

Acknowledgments		6
Introduction		7
1.	1960s: Before This Decade Is Out . . .	9
2.	1970s: Changing the Face of Space Exploration	35
3.	1980s: Reusable and Lands Like a Plane	53
4.	1990s: Forging Partnerships	69
5.	2000s: Continual Presence in Space	87
6.	2010s: Halfway to Everywhere	111

Acknowledgments

The editors and authors would like to thank the External Relations Office, the History Office, the Imagery Repository, the Scientific and Technical Information Center (STIC), and Photo Operations at NASA Johnson Space Center in Houston. Special thanks to Rebecca Wright and Jeannie Aquino—this book would not have been possible without their support. Also, thanks to Sandra Johnson, Dr. Jennifer Ross-Nazzal, Eliza Johnson, Perry Jackson, Warren Harold, Rebecca Hackler, Linda Mathews-Schmidt, Tom Sanzone, Duane Ross, Will Close, Maura White, and Clay Morgan for their help in compiling *Johnson Space Center: The First 50 Years*. The image at the bottom of page 80 appears courtesy of Peanuts Worldwide, LLC; all other images are courtesy of NASA Johnson Space Center in Houston.

INTRODUCTION

It is said that an organization is only as good as its people. Nowhere is that more true than at the NASA Johnson Space Center (JSC).

In 1961, hundreds of individuals moved from points throughout the United States to join a small task force in Houston, Texas. Together, they transformed a Texas pasture into the command post for humankind's greatest adventure. They established an avenue to space and laid the foundation for an environment of innovation, ingenuity, and integrity that continues today.

Those early pioneers and all those who carried their legacy forward shared a belief that America serves as the world leader in space exploration. This principle, evident in every program engaged at JSC, guided the workforce as it repeatedly achieved goals that extended beyond the boundaries of Earth's atmosphere.

The past five decades at the Johnson Space Center reflect an expansive list of historic accomplishments. Americans walked on the lunar surface and returned home safely thanks to JSC teams who designed the systems for space travel, provided decisive ground support, and choreographed precise recovery efforts. The success of the Apollo program came after endless hours of certifying intricate details by a workforce committed to excellence.

Building on this heritage, teams of talented and well-trained professionals at JSC expanded human spaceflight in ways that made complexity look easy. The iconic Space Shuttle, the spacecraft that launched like a rocket and landed like a plane, came from concepts created by JSC employees. They came together, exhibiting a quiet confidence, to design, to develop, to integrate, and to operate a program that would span 30 years and 135 scientific, commercial, and defense missions in space.

Technological innovations from JSC brought Americans and their former adversaries together in space—first with a brief docking of two competing spaceships then decades later for long-term residency onboard a zero-gravity facility known as the International Space Station. An unprecedented example of engineering ingenuity, this orbiting outpost provides a common ground for a multinational crew to conduct research in medicine, materials, and science to further humankind's journey into the cosmos and to help improve life on Earth.

The impact of this community also reaches far into the neighborhoods adjacent to the 1,600-acre facility, especially the educational systems. Students and faculty on all levels benefit greatly from JSC's diverse groups of current and retired scientists, doctors, engineers, managers, and astronauts. These highly qualified specialists serve as volunteers in JSC-sponsored programs that inspire the next generations.

The expertise from this workforce is legendary, being utilized for a myriad of circumstances and reflecting critical thinking with courage and conviction. Since the inception of the center, the JSC community has found brilliant and elegant solutions to difficult situations—repairing telescopes while in space, developing medical devices to save children, assembling a monumental research platform in space, and assisting with rescue efforts of those trapped underground. Even during tragic times, the JSC family moved onward with an unwavering spirit and determination to reach new heights. Their strength has taught the world to look at itself with a newer and wider perspective.

JSC's rich heritage provides a solid base for the next 50 years. As JSC continues to help humankind leave the planet, its innovative engineering services will be focused on ways to explore deep into the solar system. Teams will be making significant contributions to world-class technical research, while others forge the path for tomorrow's destinations, proving the next 50 years will be just as exciting as the last.

And, without a doubt, it will be the boundless passion and dedication of all the JSC employees that will lead the way.

One

1960s

Before This Decade Is Out . . .

On May 25, 1961, only 20 days after Alan Shepard's historic suborbital flight, Pres. John F. Kennedy addressed Congress and America, laying out an ambitious plan "of landing a man on the Moon and returning him safely to Earth" by the end of the decade. The Apollo program, one component of Kennedy's New Frontier policy, required innovation, perseverance, and commitment. The pioneers with the engineering know-how to achieve this goal included employees of the National Aeronautics and Space Administration (NASA) and its Space Task Group, both established in 1958.

The frenzy surrounding the president's deadline resulted in the establishment of a new NASA site known as the Manned Spacecraft Center (MSC). On September 19, 1961, an announcement revealed that the center would be built on more than a thousand acres of prairie 20 miles southeast of downtown Houston. The site grew rapidly in the early 1960s with buildings and personnel. Engineers and scientists, along with recent college graduates, played important roles in training astronauts, determining how to rendezvous and dock vehicles in space, engineering spacecraft, and managing missions. They used lessons learned from the Mercury and Gemini programs that set the foundation for America's future human spaceflight endeavors.

Putting an American on the Moon was a nonstop effort, and MSC's employees often worked long days and weekends. In 1967, a great tragedy struck the space program when three astronauts, the Apollo 1 crew, died during a test in their spacecraft at the Kennedy Space Center. Out of this event came a renewed sense of commitment to achieve the target date, and NASA engineers and scientists dedicated themselves to improving the safety of the vehicle.

Less than two years later, in the fall of 1968, Apollo 7, the first manned flight of Apollo, flew in Earth orbit. Apollo 8 followed in December of that year and orbited the Moon. In 1969, only eight years after Kennedy's speech, Apollo 11 made the historic first lunar landing and met the president's goal. Left behind was a plaque declaring, "We came in peace for all mankind."

Before the Manned Spacecraft Center was built, NASA's human spaceflight operations took place in leased offices around Houston. In September 1962, Pres. John F. Kennedy visited the facilities and was presented a model of the Apollo Command Module. Providing the president with details of the program's progress was Robert R. "Bob" Gilruth (above left), who served as director of the center from its inception in 1961 until 1972. Called the "father of spaceflight," Gilruth led NASA's Space Task Group, assigned to design and implement the nation's goal to put a man on the Moon. On September 12, 1962, Kennedy delivered his historic speech at Rice University, stating: "We choose to go to the Moon in this decade and do the other things, not because they are easy, but because they are hard."

From 1962 to 1964, headquarters for the NASA Manned Spacecraft Center was at 2999 South Wayside Drive in Houston, while 20 miles southeast, the permanent site facilities were being built. The former Farnsworth-Chambers Company building provided offices for the Mercury 7 astronauts and Center Director Bob Gilruth (pictured above), as well as those planning the orbital flights during the first years of the space agency. The building, constructed in 1956, is a Protected City of Houston Landmark and a Recorded Texas Historic Landmark, and it is listed in the National Register of Historic Places. The building now serves as the Houston Parks and Recreation Department headquarters.

Construction of the Manned Spacecraft Center began in April 1962 on more than a thousand acres of undeveloped land made available to the government by Rice University. The desolate site was near the shores of Clear Lake and remote farming and fishing communities in Harris County. The empty prairie became the home of the command center for the lunar landing missions and facilities for the testing, operation, and development of space exploration.

A relocation center assisted those NASA employees who transferred to Houston from the Langley Research Center in Virginia, easing their transition by providing information about housing and local services. In 1962, the area around the new space center had a limited number of places to live and shop—a marked contrast to the thriving community that exists today.

Houston embraced its designation as Space City, USA, and warmly welcomed the engineers, scientists, and technicians from all across the country as they joined NASA's workforce. Thousands of Houstonians met their new neighbors during a Fourth of July celebration in 1962 that included a downtown parade and an event at the Sam Houston Coliseum. Cowboy hats were given to the Mercury 7 astronauts, pictured above from left to right: Scott Carpenter, Gordon Cooper, John Glenn, Gus Grissom, Wally Schirra, Alan Shepard, and Deke Slayton. The next year, Houston hosted the Project Mercury Summary Conference at the Sam Houston Coliseum. More than 1,300 people attended to learn details of the Mercury program, which had lasted 55 months.

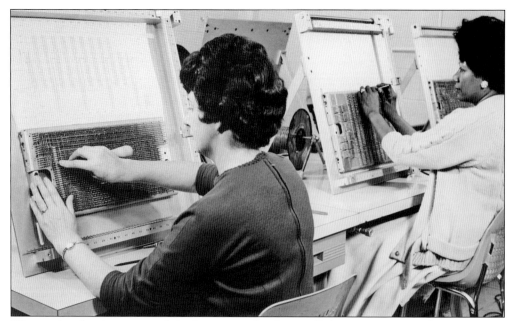

Women played an important role in the success of early spaceflight programs, not only as secretaries but also as machine operators. Before the days of modern computers, the commands and data that controlled spacecraft operations were sent through electrical circuit boards that required manual input by the crew. This photograph from 1964 shows women using specialized tools to perform adjustments to the circuitry.

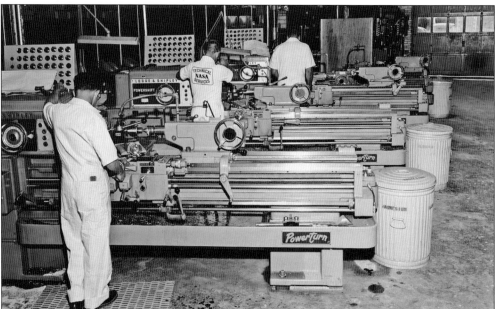

By 1962, the initial Technical Services Division personnel had relocated to Houston from Langley Research Center in Virginia. A vacant Canada Dry bottling plant became the new home for this group, and from their tools and machinery, they crafted everything from life-size spacecraft mock-ups to small-scale models for exhibits and displays. Technicians also installed and serviced equipment for the scientific and technical shops throughout the Manned Spacecraft Center.

On February 20, 1964, the first 280 employees began the slow process of relocating offices, laboratories, machinery, and equipment to the newly built facilities in Clear Lake. Over the next five months, more than 2,700 people moved from temporary locations throughout Houston to the new center. Pictured are Building 13, the Systems Evaluation Lab, and Building 15, the Instrument and Electronics Lab.

Crews trained with Gemini static articles and boilerplate capsules in the 55,080-gallon water tank constructed in Hangar 135 at nearby Ellington Air Force Base. Water egress and recovery exercises allowed the crews and recovery personnel to simulate the activities necessary after splashdown of the Gemini spacecraft. The 16-foot-deep tank had four underwater viewing ports used for photographic documentation and test observation, and in 1966, it was moved onsite to Building 260.

MSC's Landing and Recovery Division used a World War II–era tank landing craft to train astronauts for post-splashdown ocean recovery operations and spacecraft qualification tests. Named the MV *Retriever*, the sides of the 114-foot vessel's midsection were cut down, a new bridge built, and a hoist added for NASA use. Training and testing with the vessel occurred primarily in Galveston Bay and the Gulf of Mexico.

Photographic documentation of spaceflight test activities allowed engineers and scientists to ensure the safest possible environment for the astronauts while still accomplishing mission objectives. High-speed, long-lens cameras were used to cover testing in Galveston Bay as a Gemini boilerplate capsule was dropped from an Air Force C-119 aircraft. Test instruments with sensors attached to the capsule measured the impact forces of a water landing compared to land landings.

Emergency training for astronauts included escape methods under every conceivable occurrence. The Gemini spacecraft escape system used ejection seats similar to those in jet fighter aircraft to allow astronauts to break free from the capsule during a low-altitude launch abort or landing emergency descent. During training, astronauts wearing Gemini-type pressure suits and parachutes were towed by a powerboat using a specially designed parachute connected to a 600-foot towrope. After ascending to an altitude of up to 400 feet, the astronauts dropped from the canopy for a free descent by parachute into Galveston Bay, where they were picked up by scuba divers and training personnel (pictured below). Water survival training included pressure suit flotation, underwater egress from the spacecraft, helicopter pick-up, life raft boarding, parachute drag escape, and shroud line disentanglement.

Pres. Lyndon B. Johnson (center) visited the Manned Spacecraft Center for the first time on June 11, 1965, and was briefed on the success of the Gemini 4 mission, which included Ed White's historic 22-minute spacewalk. White (left) showed the president the hand-held self-maneuvering unit or "zip" gun, designed by MSC engineers, that allowed him limited movement around the Gemini capsule.

One of the Gemini program's main objectives was orbital rendezvous. On December 15, 1965, Gemini 6 and 7 achieved that goal when the two crewed capsules flew in formation within one foot of each other and remained in close proximity for five hours. The Mission Control Center in Houston, including Chris Kraft (left), Gordon Cooper (center), and Bob Gilruth (right), reveled in the success of the joint mission with their traditional cigars.

During a pre-mission press conference in July 1966, the Gemini 10 commander used a chalkboard to illustrate a tethered spacewalk planned for the upcoming flight that also included the first double rendezvous in space. Earlier in the year, the Gemini 8 mission had accomplished the first ever orbital docking when the capsule successfully docked with its pre-launched Agena Target Vehicle (ATV). However, a serious thruster malfunction forced the mission to end with an emergency landing only 10 hours after launch and the abandoned ATV later served as the second rendezvous target for the Gemini 10 crew. Also during this historic event, the Gemini 10 pilot became the first human to "dock" with another spacecraft in orbit. While at the end of a 50-foot tether, he retrieved a micrometeorite detector from the side of the drifting ATV during a 49-minute spacewalk.

Water egress procedure training was crucial for the crews assigned to the Apollo spaceflight missions, and astronauts had to be prepared for a variety of environmental conditions after splashdown. Trainers took advantage of every available body of water at their disposal, including nearby Ellington Air Force Base swimming pools, a specially built tank at the base facility, and Galveston Bay. Aided by rescue divers, the Apollo 1 crew of Ed White, Roger Chaffee, and Gus Grissom practiced exiting the capsule in one of Ellington's pools on a sunny day in 1966. Their mission was to be the first manned Apollo flight in February 1967. In January of that year, while their Command Module was mounted on top of the Saturn 1B on the launchpad in Florida, a simulation was conducted to reproduce the launch environment. Tragically, a fire broke out in the sealed crew cabin, trapping the three astronauts inside. The loss of the crew sent shockwaves through NASA and the world. It would be almost two years before America returned to manned spaceflight.

The Manned Spacecraft Center hosted a number of diverse international delegations interested in the pioneering efforts of early spaceflight. In this 1966 photograph, Max Faget explains a model of the Apollo spacecraft to members of the Korean National Assembly. Faget served as the director of engineering and development and as the lead spacecraft design engineer for the Mercury, Gemini, Apollo, and Space Shuttle programs.

The Space Environmental Simulation Laboratory consists of two separate vacuum chambers that recreate the extreme conditions of space. With interior dimensions of 90 feet in height and 55 feet in diameter, Chamber A accommodated the Apollo Command and Service Module with room to spare. As the floor rotated, the spacecraft components were tested against the simulated pressure and temperature ranges of space in what engineers nicknamed the "barbecue roll."

To determine the effects of gravity, or "G," forces on the human body, engineers added a centrifuge in Building 29 to the Apollo testing protocol. Test subjects could be whirled around a circular course at 24 revolutions per minute and speeds up to 88 miles per hour to evaluate and qualify lunar spacecraft equipment. The mechanism also allowed astronauts to become familiar with the Apollo launch and reentry profile accelerations and the G forces encountered in spaceflight. For the first man-rating tests on the newly built centrifuge, System Test Branch personnel used a modified Gemini seat mounted upright in a swing cradle below the arm (above). After the initial training period, test subjects, and later astronauts, rode the centrifuge side by side strapped into couches inside a three-ton metal ball, or gondola, at the end of the 50-foot arm.

Development of spacecraft required unique laboratories, including an anechoic chamber located in Building 14. Designed to accommodate large test articles, the echoless environment replicates space by absorbing radio and radiation emissions. Foam pyramids absorb stray radiation during spacecraft antenna radiation pattern tests. The chamber was used numerous times by engineers to prepare the Apollo Command Module for its missions to the Moon.

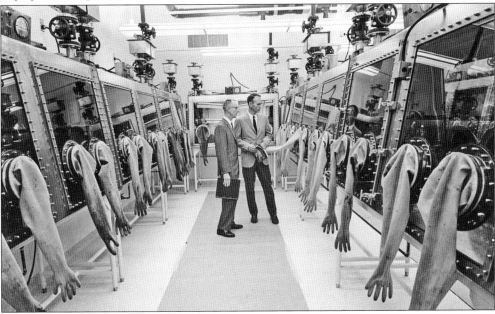

The Lunar Receiving Laboratory, completed in September 1967, housed geological and biological testing laboratories and served as the Apollo astronaut quarantine facility after the first three successful landing missions returned from the Moon. The glove box system (pictured above) allowed technicians to manually manipulate lunar sample return containers in vacuum chamber cabinets using impermeable gloves built into the chamber wall and tools stored within the primary biological barrier.

Vibration-acoustic tests were conducted on the Apollo spacecraft "stack" in Building 49. To confirm structural strength and design models, the stack was mounted on a series of vibration shakers located on the floor. The spacecraft was also subjected to airborne forces emanating from the firings of the rocket engines and the aerodynamic noise and shockwaves associated with the vehicle's passage along the liftoff flight path.

Astronauts in the Lunar Module needed a reliable system to ascend from the lunar surface and rendezvous with the orbiting Command Module. After docking, the Command Module would then return the astronauts, along with the lunar samples they collected, safely to Earth. Engineers in Building 353 conducted a series of tests at sea level conditions on this vital ascent propulsion system.

In addition to collecting samples to bring back to Earth, Apollo 11 astronauts left behind the Early Apollo Scientific Experiment Package on the surface of the Moon with instruments to monitor lunar conditions. Astronauts trained with experiments while wearing their "backpacks," or Primary Life Support Systems, developed by MSC engineering teams.

The 2TV-1 spacecraft was built to the same specifications, of the same materials, and with nearly all the flight-qualified equipment aboard the Apollo capsule. In mid-1968, three astronauts spent seven days in this vehicle placed in the Space Environment Simulation Laboratory Chamber A. The crew and the 2TV-1 were monitored while undergoing conditions and temperatures expected during spaceflight to verify the spacecraft structure, pressure vessel, and the environmental control system.

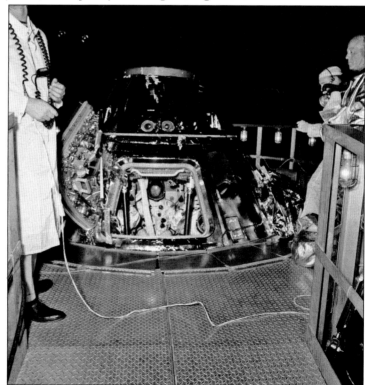

Just weeks before the first manned Apollo mission, Director of Flight Operations Chris Kraft (standing) explained the Mission Control Center's functions and procedures to Prince Rainier and Princess Grace of Monaco. The royal couple, seated at center in the Mission Operations Control Room viewing area, toured the Manned Spacecraft Center and learned about the planned Moon landings from MSC director Bob Gilruth (front row) and his staff. Throughout the years, dignitaries and international officials frequently visited the viewing room, which allowed a behind-the-scenes glimpse of flight control activities. The flight control operations resulting from the Mercury, Gemini, and Apollo programs became a catalyst that changed the world of real-time and voice communications. Cable networking and interfacing inspired by spaceflight requirements helped to connect the world together in ways only dreamed of before the human spaceflight program.

More than 2,000 people converged at Ellington Air Force Base in the middle of the night to welcome home the Apollo 8 crew. The groundbreaking mission took humans closer to another celestial body than ever before, within approximately 68 miles of the lunar surface. After 10 orbits around the Moon, and with no contact with Mission Control, the crew performed a critical maneuver on the dark side of the Moon before starting their journey back home. The following mission, Apollo 9, tested the Lunar Module and other hardware to check out rendezvous and docking procedures. A spacewalk performed during the flight certified the new Apollo spacesuit, the first to have its own life-support system independent of the spacecraft. Below, the crew of Apollo 9 returned to Houston to the waiting arms of their joyous families.

Apollo 11 commander Neil Armstrong (left) and Lunar Module pilot Buzz Aldrin practiced deploying the Early Apollo Scientific Experiments Package during a training session in Building 5. The astronauts learned to operate cameras supported by brackets mounted on the front of their spacesuits, take photographs without the aid of a viewfinder, and manipulate the controls and trigger fitted underneath the camera.

Apollo 11 Command Module pilot Michael Collins was responsible for rendezvous and docking with the Lunar Module once it returned from the Moon. He spent many hours inside the spacecraft mock-up, sometimes without his fellow crewmen, simulating these critical procedures and preparing for any possible outcome or emergency.

The final phase of every successful Apollo Moon landing was manually piloted by the mission commander. To prepare for that critical phase of descent in the Lunar Module, astronauts trained at Ellington Air Force Base with an open framework Lunar Landing Training Vehicle (LLTV). A 4,200-pound-thrust turbofan engine, modified for vertical use and mounted behind the cockpit, counteracted five-sixths of the vehicle's weight, putting it in a simulated lunar one-sixth gravity. Nicknamed the "flying bedstead," the LLTV could reach up to 800 feet in altitude. By using thrusters similar to the Lunar Module and vertically mounted hydrogen peroxide lift rockets for control, the astronauts practiced maneuvering and landing on the lunar surface while still on Earth. Two of the three LLTVs built were destroyed in training accidents; the pilots ejected safely.

On July 20, 1969, eight years after Pres. John F. Kennedy's announcement, the Apollo 11 Lunar Module touched down on the Sea of Tranquility. While the rest of the world celebrated when hearing the words, "Houston, Tranquility Base here. The *Eagle* has landed," flight controllers in the Building 30 Mission Operations Control Room (MOCR) stayed focused on the mission. Seen below, supporting the MOCR was a pool of systems engineers and experts, in the Building 45 Mission Evaluation Room, who closely followed operations on 19-inch monitors linked to the various MOCR flight control displays. Engineers recorded the status of measurements with Polaroid instant cameras. They used audio headsets to listen to the transmissions from space and various flight controller "loops" by using a pushbutton audio channel selector box.

On July 24, 1969, the Apollo 11 crew returned safely to Earth, completing the goals of the first manned mission to the Moon. On the lunar surface, they had left a number of items, including a plaque stating, "Here men from planet Earth first set foot upon the Moon. July 1969 A.D. We came in peace for all mankind." Hailed as possibly the greatest technological achievement of the 20th century, thousands of people throughout NASA and the United States had contributed to the historic effort. The success of this mission came as a result of years of planning and testing by a dedicated workforce. Celebrating this achievement with the rest of humankind were the flight controllers in the Mission Operations Control Room (above), who had controlled the Apollo 11 mission from liftoff to splashdown. The average age of the group was only 26 years old.

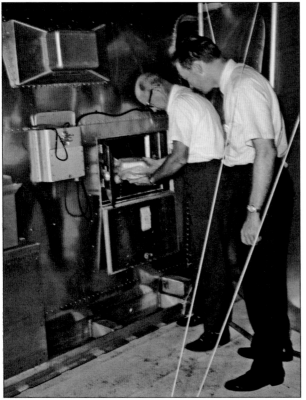

Astronauts returning from the first three Apollo lunar missions were immediately quarantined after splashdown in the Mobile Quarantine Facility (MQF). Via air transports, the MQF, a converted Airstream trailer, was moved from the US Navy aircraft carrier to Ellington Air Force Base, then towed to the space center, where it was docked in Building 37, the Lunar Receiving Laboratory (above). Sequestered in the facility, the astronauts utilized a kitchen, sleeping and living areas, and space to communicate with their families and NASA personnel while maintaining separation. Also housed in the laboratory were materials brought from the lunar surface. Pictured at left, to prevent contamination, items including the canisters of film were passed through the airlock to be unpacked, sorted, weighed, and labeled by scientists in Building 37.

Lunar samples collected by Apollo 11 astronauts were flown to Houston and transferred directly to the Lunar Receiving Laboratory. Each box contained 50 rocks, samples of lunar "soil," and two core tubes of material from below the Moon's surface. On hand for the arrival of the first geologic samples was Center Director Bob Gilruth (right).

Workers loaded the Apollo 11 Command Module aboard a Super Guppy aircraft at Ellington Air Force Base. The spacecraft, released from its postflight quarantine at the Manned Spacecraft Center, was then shipped to the manufacturer in California for evaluation of the capsule's heat shield, damaged by the extreme heat of reentry.

The city of Houston welcomed the Apollo 11 crew back home with a parade through the streets of downtown. On August 16, 1969, an estimated 300,000 people filled the city sidewalks and celebrated the three astronauts' achievement. Continuing the festivities, 45,000 people attended a celebrity-filled event held at the Astrodome later that day.

In the years following Yuri Gagarin's flight, astronauts and cosmonauts had met a number of times, but their meetings had been shadowed by the Cold War. In October 1969, Russian cosmonauts (right) toured the Manned Spacecraft Center for the first time during a two-week trip to the United States. Less than 10 years later, space travelers from both countries worked together on a joint mission called the Apollo-Soyuz Test Project.

Two
1970s
Changing the Face of Space Exploration

Landing a man on the Moon was a major achievement and a source of great pride for Americans, but interest in mankind's greatest adventure began to dwindle after the first few lunar landings and the dramatic rescue of the Apollo 13 astronauts. Just a few months after these great technological feats, cuts to NASA's budget forced the cancellation of the last two scheduled Apollo flights. As the lunar expeditions came to an end, and facing the harsh economic realities of the 1970s, the Manned Spacecraft Center began reducing its workforce.

NASA continued to set challenging goals for future human spaceflight programs, which included the world's first reusable spacecraft that would provide routine access to Earth orbit, the Space Shuttle. Engineers in Houston dedicated much of the 1970s to designing, developing, and testing the numerous systems of this vehicle, while managing the Shuttle Program Office for the agency.

NASA promoted cost-effective solutions to the decade's budget woes, including Skylab, America's first space station, which relied on already proven Apollo hardware. Astronauts flew three long-duration expeditions where they participated in medical experiments and gathered data about Earth and solar activities.

This decade was also a period of Cold War détente and social change for the center. The United States and the Soviet Union's competition to reach the Moon morphed into fledgling cooperation as the two countries worked together on their first joint mission, the Apollo-Soyuz Test Project. In 1973, the Manned Spacecraft Center was renamed in honor of the late Pres. Lyndon B. Johnson, and it would henceforth be known as the Johnson Space Center, or JSC. Throughout the decade, women and minorities became increasingly more visible at the center and within previously all-male offices like the astronaut corps.

A time of transition, the 1970s set the foundation for the center's future endeavors.

During the 1960s, acreage for the new federal spaceflight laboratory quickly transformed into the Manned Spacecraft Center (seen above in the early 1970s). Numerous state and national contractors and suppliers participated in the construction, and the pioneering concept of using prefabricated exterior building panels, which has since become common practice, sped the construction process. The Texas Legislature authorized and funded NASA Road 1 (bottom right corner), a unique and distinctive highway category without precedent that propelled expansion and growth to the areas around the space center. Dominating the center's new landscape was the nine-story project-management headquarters. Nearby were computer buildings, spacecraft-environment chambers, astronaut training facilities, and the Mission Control Center. By the early 1970s, the center had been touched by both tragedy and triumph and was in the process of preparing for the next phase of exploration and discovery.

Just 56 hours after launch, Apollo 13 crewmembers heard a loud bang when an explosion onboard their spacecraft caused the cancellation of their mission to the lunar surface and placed them in peril. In the Mission Operations Control Room (above), flight controllers and their dedicated team members worked together nonstop for the next four days, making decision after decision in formulating a plan to bring the astronauts safely home. On April 17, 1970, the astronauts jettisoned the Lunar Module and in the Command Module made their reentry through the atmosphere, landing in the South Pacific. Below, celebrations erupted in Mission Control when the recovered crew was seen aboard the USS *Iwo Jima*, ending the most severe in-flight emergency faced thus far in human spaceflight.

Between 1969 and 1972, six Apollo spaceflight missions brought back 842 pounds of rocks and soil from the surface of the Moon. The specially built Lunar Receiving Laboratory (LRL) became the home for the extraterrestrial materials, as well as the temporary home for the first three successful Apollo crews while they were quarantined for a period of time after landing. One of the most complex laboratories in American history, the LRL brought prominent biological and geological scientists from the around the world to Houston during the lunar program. Wearing special germ-free clothing, Center Director Bob Gilruth (left, at right) looks on as samples collected during one of the Apollo missions are inspected. Below, contents of the second Apollo 11 sample return container are examined by LRL scientists using impermeable gloves and a vacuum chamber cabinet.

The iconic Earthrise (above) from Apollo 8 provided the first image of the world taken by humans. The first people to see the picture worked in the Photographic Technology Laboratory at the Manned Spacecraft Center. These technicians processed thousands of photographs from the film rolls brought back to Houston by astronauts. During the first missions, handheld cameras with 80- and 250-millimeter lenses captured the unparalleled scientific observations. Advising and training the crewmembers was a team of experts including geologists, astronomers, and photographers. Added to the later missions were cameras mounted on the spacecraft modules or instruments that were operated remotely by astronauts while in orbit. Below, when the Apollo program ended with Apollo 17, astronauts had photographed nearly 20 percent of the lunar surface in detail and provided a visual understanding of the Moon and Earth.

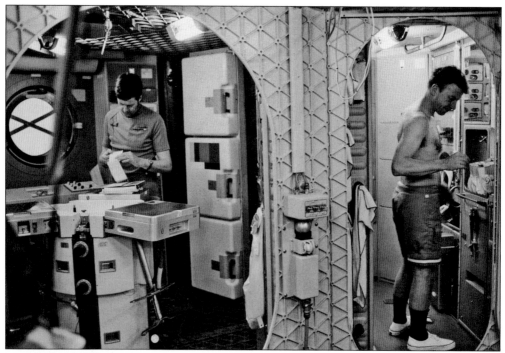

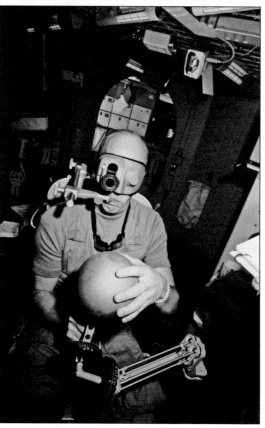

Skylab was launched into Earth orbit by a Saturn V rocket on May 14, 1973, to provide humans an environment to live and work in space for extended periods. America's first space station orbited the planet from 1973 to 1979 and included a workshop and solar observatory. Prior to launch, crews simulated mission tasks in the trainer at JSC (above) and checked out experiments that would be used while in space (below). Three separate crews occupied the Skylab workshop for a total of 171 days and 13 hours. Astronauts conducted nearly 300 scientific and technical experiments, including medical experiments on humans' adaptability to zero gravity, solar experiments, and detailed Earth resources experiments. The unmanned Skylab burned up on reentry in 1979, scattering debris over the Indian Ocean and Western Australia.

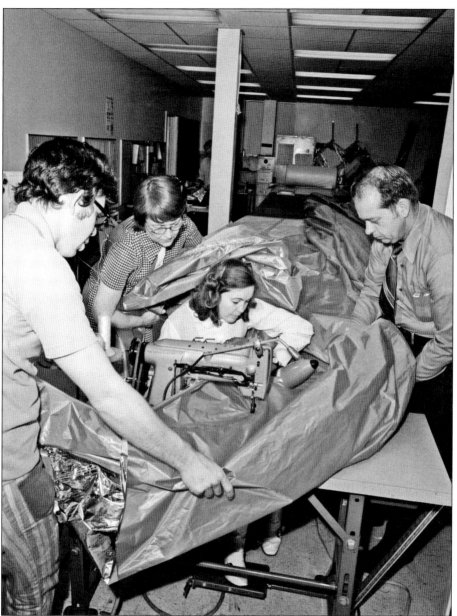

One of Skylab 1's two main solar panels, vital to the space station's success, was critically damaged during launch. Internal temperatures onboard reached 126 degrees Fahrenheit, potentially jeopardizing the workshop's experiments and mission. Immediately, NASA began to seek solutions from across the agency. The design from the Houston team was selected as the best answer—a sunshade to protect the orbital workshop from the Sun's excessive heat. Above, technicians help a seamstress feed material for the three-layered sunshade through a sewing machine. Developed in just one week, the temporary solar shield was carried with the crew of Skylab 2 when they launched on May 25, 1973. The astronauts repaired the workshop by setting up the umbrella-like canopy over the spacecraft's exterior and stretching out Skylab's second solar panel. By June 4, the workshop was fully operational, and it successfully served the crew until their departure on June 22, 1973.

In August 1973, on what would have been former Pres. Lyndon B. Johnson's 65th birthday, his widow, Lady Bird Johnson, and her family joined NASA officials to formally dedicate the center in honor of her late husband. The former first lady discussed Johnson's dedication to space exploration, describing him as a man who "from the beginning believed in more than just the conquest of space, who wanted to bridle this great force for the cause of peace, and who believed that the infinity of space could provide a common ground for all nations to work peacefully together—putting aside the differences, suspicions, and conflicts of this planet." The sign on NASA Road 1 below marks the entrance to the Johnson Space Center, a site transformed from a "Texas pasture into the command post for humankind's greatest adventure."

Artist Robert "Bob" McCall (right) painted *The Next Giant Step* in 1979 while visitors watched. The 72-foot-by-16-foot acrylic painting on canvas commemorates the heroism and courage of spaceflight pioneers and includes the faces of many of the center's employees. Located in the lobby of Building 2 just outside of the Olin Teague Auditorium, the mural depicts America's journey of space exploration and includes visions of future travel.

Building 2 housed the visitor's center for 30 years until 1992, when Space Center Houston was constructed as a separate facility adjacent to the site. Before then, tourists were able to drive into the center, park, and walk through Building 2 to marvel at a selection of artifacts from the Mercury, Gemini, and Apollo eras. Thousands of schoolchildren each year enjoyed field trips to the center.

Pres. Richard M. Nixon addressed employees and visitors who gathered outside Building 2 in March 1974 to hear him pay respects to the "great American scientific, mechanical, and clerical communities that made it possible for America" to be a leader in space travel. During the ceremony, the president presented NASA Distinguished Service Medals to the crewmembers of Skylab 3.

The final mission of the Apollo program in July 1975 was also an historic first, as three US astronauts and two Soviet cosmonauts successfully docked their respective vehicles in space for the joint Apollo-Soyuz Test Project. Training simulations took place in Building 13, where both American and Soviet engineers tested the compatibility of docking mechanisms. The test control room can be seen on the right.

One of the most difficult problems to overcome for the Apollo-Soyuz Test Project was that of language differences. To address this problem, the US astronauts learned Russian (shown above); their Soviet counterparts studied English. During the mission, all communications among the five crewmembers were made in the language of the listener, with the Americans speaking Russian to the Soviet crew and the Soviet crew speaking English to the Americans.

On July 16, 1975, the world watched as the Apollo-Soyuz Test Project space travelers shared a historic handshake in docked spacecraft. Providing coverage of this unprecedented event were members of the international press corps gathered at the center. Throughout the past 50 years, the Public Affairs Office has assisted US and foreign reporters representing television networks, wire services, radio stations, newspapers, magazines, scientific and educational publications.

The Saturn V rocket arrived in Houston in the fall of 1977 to become part of a permanent visitor's exhibit at JSC. Parts came from all across the country to form a display with the first stage brought in by barge from Louisiana to Galveston Bay and from there to a Clear Lake dock near JSC. Below, the Saturn V became the centerpiece of Rocket Park, located near the main entrance of the center. Designated a National Historic Landmark, this Saturn V was assembled from flight-ready stages and spacecraft abandoned after cancellation of the Apollo and Skylab programs.

Thirteen Saturn V rockets launched between 1967 and 1973, and nine took humans to the Moon. The last Saturn V launched Skylab into orbit. Built in the late 1960s, the Saturn V extended 363 feet long and was capable of generating 7.5 million pounds of thrust. This magnificent rocket remains the largest, most powerful American launch vehicle ever built. Three Saturn V rockets still exist, but only the one at JSC consists of stages actually intended for spaceflight. After being on its side and subject to the Texas Gulf Coast weather for more than 20 years, the rocket suffered deterioration, corrosion, paint failure, moisture infiltration, and structural failures. In 1999, the Smithsonian Institution, which owns the Saturn V and loans it back to NASA for display, began fundraising for the rocket's restoration. The Save America's Treasures program, National Park Service, and the National Trust for Historic Preservation provided half of the funds required for its preservation. A facility was completed in 2007 to enclose this symbol of one of America's greatest technological accomplishments.

In 1972, Pres. Richard M. Nixon approved the Space Shuttle, and MSC's engineers became responsible for designing, developing, and testing the Orbiter. Testing of the reusable vehicle was extensive. In order to prove the reliability, safety, or quality of the Shuttle's subsystems, engineers tested parts, materials, and entire subsystems. Houston's state-of-the-art facilities, less than a decade old, were heavily involved in this effort. The anechoic chamber, featured here, conducted antenna testing on a 1/10th-scale model of the vehicle. Data gathered during runs at the center's other major labs, such as the Vibration and Acoustic Test Facility, the Shuttle Avionics and Integration Laboratory, and the Space Environment Simulation Lab, demonstrated the viability of the spacecraft's systems. If a part, material, or subsystem failed, engineers had to replace inferior materials with new substances or redesign entire subsystems.

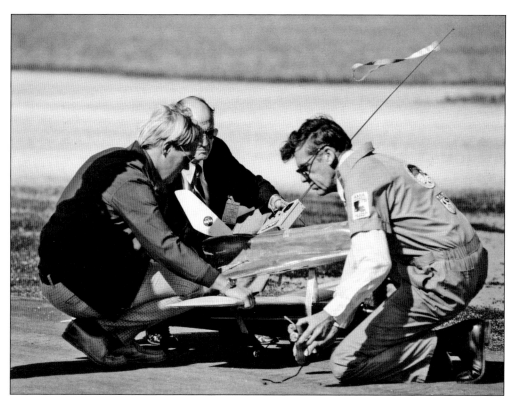

An innovative idea from a longtime JSC engineer and model builder resulted in the development of the Shuttle Carrier Aircraft, the cross-country ferry transport for the Orbiter. The design was proven in 1977 in the open field of the Building 14 Antenna Test Range by using 1/40th-scale models of the Orbiter and the Boeing 747. Engineers studied the Orbiter control characteristics and separation dynamics while watching the model ride piggyback on top of the small plane, then separate in mid-air and land. Although many crashes occurred before the concept was proven, the experimental flights served as the precursor to the Approach and Landing Tests, where a full-size aircraft and the Space Shuttle *Enterprise* were used to confirm the design as a successful process.

The face of American space exploration changed in 1978, when NASA selected the first class of astronauts that included women and minorities. This groundbreaking eighth class, known as the "Thirty-Five New Guys," also introduced another form of diversity to the astronaut corps. In addition to the military pilots who would be responsible for commanding the Shuttle and returning it safely to Earth, this group included 20 scientists, doctors, and engineers with advanced degrees in their fields to serve as mission specialists and conduct experiments in space. After their careers as astronauts, several of the members from this class went on to hold leadership positions at NASA and in the aerospace industry. In 1992, the agency expanded its astronaut corps to include international members from countries such as Canada, Japan, and France. Today, JSC continues its role as the NASA center responsible for directing the astronaut selection process and as the primary site for crew training.

Shuttle astronauts trained for zero-gravity conditions of space aboard a modified KC-135 aircraft. The so-called "Vomit Comet" allowed passengers to experience 25-second intervals of weightlessness as it flew up and down in a series of parabolic maneuvers. During these training sessions, scientists and engineers conducted medical and motion sickness experiments on the astronauts, while also developing and verifying space hardware. The Water Immersion Facility (bottom) offered another way for astronauts to practice working in the weightlessness of space. The 33-foot-by-78-foot, 25-foot-deep pool contained submerged mock-ups of the Orbiter and other mission equipment in a neutrally buoyant environment. Specially trained divers helped the astronauts complete their underwater training tasks, while technicians and medical personnel assisted on the surface.

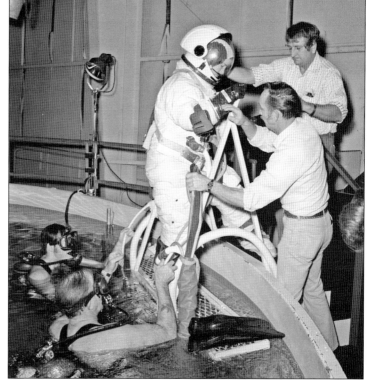

Nearly a quarter of a million people went to see *Enterprise*, the first Space Shuttle Orbiter, in March 1978 at Ellington Field, Houston. The spacecraft, atop a modified 747 airplane, was on its way to Marshall Space Flight Center, Alabama, for ground vibration tests. From February through November 1977, *Enterprise* had been used for the Shuttle Approach and Landing Test Program at NASA's Dryden Flight Research Center in California. Ground and flight tests demonstrated that the Orbiter could fly in the atmosphere and glide safely to a landing without jet engines. Tests were also conducted while the *Enterprise* was mated atop the 747 (Shuttle Carrier Aircraft) to determine capability for handling and control. Although it was originally named *Constitution*, fans of the 1960s science fiction television show *Star Trek* initiated a letter-writing campaign to rename the vehicle after the fictional starship. Designated by NASA as OV-101, the vehicle rolled out of the Palmdale, California, assembly facility on September 17, 1976. Four months later, it was transported 36 miles overland to Dryden to begin its legacy.

Three
1980s
Reusable and Lands Like a Plane

JSC engineers dedicated much of the 1970s to designing, developing, and testing the world's first reusable spacecraft, and by the 1980s, the Space Shuttle was ready for flight. On April 12, 1981, *Columbia* launched on its inaugural mission, a two-day test flight, from Kennedy Space Center in Florida. After the fourth and final orbital test flight, Pres. Ronald Reagan declared the vehicle fully operational. The Shuttle fleet went on to fly until July 2011.

Flying the Shuttle became a business for NASA, with customers from commercial industry and the Department of Defense (DoD). A significant number of the 20 missions flown between November 1982 and mid-January 1986 deployed communication satellites from the Shuttle's payload bay. Two classified DoD missions flew in 1985 in a marked departure from NASA's open disclosure policy.

Shuttle flight crews reflected the changing nature of spaceflight in the 1980s. Women and minorities selected in 1978 and 1980 altered the look of crews. Along with professional astronauts, there was a new group named payload specialists—space fliers from industry, the military, other nations, or Congress assigned to the mission for a designated purpose.

As in decades past, tragedy in human spaceflight accompanied the triumphs. On January 28, 1986, *Challenger* and the most diverse crew to date were lost over the Atlantic Ocean. The world was traumatized by the tragedy. All flights were abruptly postponed as the agency worked to correct the flaws identified by the Presidential Commission on the Space Shuttle *Challenger* Accident. In total, NASA made 76 changes to the vehicle, including the installation of new brakes and the addition of a crew escape system. In September 1988, the Shuttle returned to flying, restoring workforce confidence in the vehicle and launch system and helping to heal the wounds of the loss.

In the 1980s, NASA also began laying the foundations for the International Space Station. JSC and the Shuttle would play a critical role in the construction of this outpost in space, coordinating with partners from around the globe.

Crews prepared for missions at the Building 9 Shuttle Mock-up and Integration Laboratory. Using the Full Fuselage Trainer (left) and Crew Compartment Trainers, astronauts familiarized themselves with the Shuttle's systems and emergency escape procedures. The sectioned-off Manipulator Development Facility helped astronauts practice operating the robotic arm. The building was later known as the Space Vehicle Mock-up Facility.

In 1980, the Sheet Metal and Model Branch of JSC's Technical Services Division built the Space Shuttle Full Fuselage Trainer out of wood (section shown here). This trainer was just one component needed to prepare for the upcoming Space Transportation System (STS) missions. The full-scale, wingless mock-up included many of the Shuttle's key features, such as the cockpit, middeck, and payload bay.

In preparation for on-orbit contingencies and extravehicular activities, trainers installed a mock-up of the Orbiter's payload bay inside the newly built Weightless Environment Training Facility (WETF) in Building 29. Astronauts utilized the WETF, which simulated the effects of zero gravity, to practice the choreographed maneuvers they would complete on a spacewalk. Crews from a host of flights, including the first two Hubble Space Telescope servicing missions, trained in this facility.

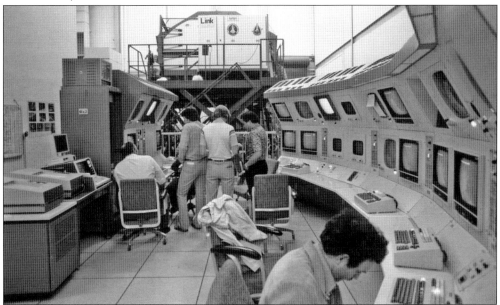

The Shuttle Mission Simulator, which consisted of fixed- and motion-based simulators, trained flight crews. Instructors sitting at consoles scripted scenarios to simulate critical system failures that astronauts could encounter on lift-off, entry, and in orbit. A commander, pilot, and flight engineer practiced ascent and entries on the motion-based simulator, which replicated the sights and sounds of launch and entry. Astronauts practiced on-orbit tasks in the fixed-based simulator.

STS-1 launched a new era for the United States, returning the nation to space after the close out of the Apollo program in 1975. Called the boldest test flight in history, *Columbia* launched on April 12, 1981. This historic mission marked the first test flight of the Shuttle into space and the only US manned test flight of a new spacecraft system. Aboard *Columbia* for the inaugural flight were, from left to right, Pilot Bob Crippen and Commander John Young; they are pictured with Dick Truly and Joe Engle, the STS-2 crew. In the bottom photograph, flight controllers work diligently in the Mission Operations Control Room during the STS-1 mission, which lasted 54.5 hours. The crew landed *Columbia* at Edwards Air Force Base in California on April 14, 1981.

The Guppy aircraft was developed in 1962 to satisfy NASA's requirement of ferrying large, outsized cargo components. The unique Pregnant Guppy and Super Guppy played a significant role in meeting launch schedules and transporting mammoth components, flying over two million miles in support of the Gemini, Apollo, and Skylab programs. The aircraft allowed airlifting of equipment that would otherwise have been sent by ship, taking weeks or months longer and delaying program timelines. In the 1980s, the Super Guppy (above, flying over JSC) continued to support the Space Shuttle program. Loading the Guppy aircraft was simple and efficient because of the hinged nose, which opens 110 degrees for cargo loading, and a system of rails and lock pins in the cargo compartment. With an interior diameter of 25 feet and a cargo compartment length of 111 feet, the Guppy could carry over 52,000 pounds of payload. Cruising at 25,000 feet and 290 miles per hour, the aircraft provided fast, economical transport over the years.

Center Director Chris Kraft describes the functions of the Mission Operations Control Room to Pres. Ronald Reagan, who visited the center while STS-2 was in orbit. Reagan spoke with the crew: "I'm sure you know how proud everyone down here is and how this whole nation, and I'm sure the world, but certainly America, has got its eyes and its heart on you."

On March 9, 1982, JSC recognized its 15 millionth recorded visitor. Since June 1964, the space center has welcomed the public to visit the site, Mission Control, and the astronaut training facilities. In 1981, when America returned to flying in space after the Apollo program, 1.5 million people toured the facilities and onsite visitor's center.

Flight directors huddle with an astronaut in the Mission Operations Control Room to discuss reentry for STS-3. NASA's primary landing site for the Orbital Flight Tests was Rogers Dry Lake at Edwards Air Force Base. The lakebed's longest runway (7.5 miles) offered astronauts plenty of space to bring the vehicle to a stop, but heavy rains made the runway muddy in 1982, forcing NASA to change the STS-3 landing site to Northrup Strip at White Sands, New Mexico (below). On March 30, 1982, *Columbia* safely landed on the strip, which was later renamed the White Sands Space Harbor. This was the only time a Shuttle landed in New Mexico.

During the STS-51D mission in April 1985, a satellite failed to ignite after deployment, so Mission Control designed a plan to save it. Detailed instructions were written for the crew to use after *Discovery* had completed a rendezvous with the malfunctioning satellite—but prior to the attempt in orbit, the sequence was tested in Building 9 in the Manipulator Development Facility (pictured below). First, a "flyswatter" device was constructed from materials available on the vehicle, then attached to the end of the arm (pictured above). Using this improvised tool, an astronaut triggered the ignition switch on a mock-up of the satellite suspended above the cargo bay. The process worked on the ground, but unfortunately, the satellite would not start in orbit. A few months later, the STS-51I crew repaired the satellite.

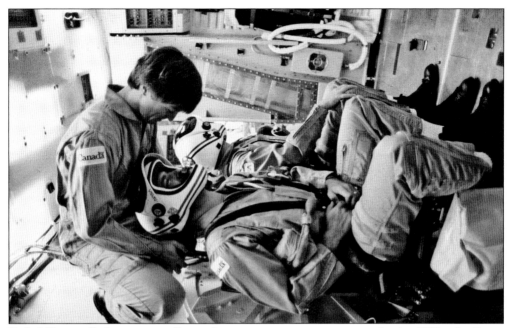

Early US spaceflight programs flew only career astronauts into space. In the 1980s, the Space Shuttle program extended spaceflight opportunities to nonprofessional astronauts, including individuals from industry, the Department of Defense, America's partners in space exploration such as Canada, and members of Congress. Known as payload specialists, these members of the crew were responsible for a specific payload or experiment.

Prior to STS-41C, in April 1984, a reporter (left) examined the tools to be used on orbit to repair the Solar Maximum satellite. This mission marked the first use of the Manned Maneuvering Unit (MMU), giving astronauts the ability to capture, retrieve, and repair communications satellites. Two months earlier, the MMU had been tested in orbit, proving that the untethered propulsion unit could take crew members up to 320 feet from the Orbiter.

Vice Pres. George H.W. Bush visited JSC in April 1983. During the tour of Mission Control, Center Director Gerald Griffin (right) shared information with the vice president and NASA administrator James Beggs, including details of the STS-6 flight, which was currently in orbit. While president, Bush again visited the center.

Jesse Moore became director of the Johnson Space Center in 1986, only five days before the tragic loss of the Space Shuttle *Challenger*. After the accident, Moore addressed the devastated employees, promising, "We will continue" to fly into space "with a renewed spirit and a strengthened sense of purpose."

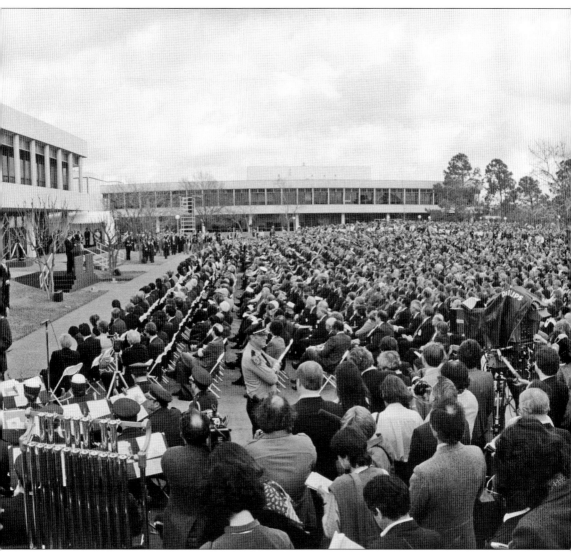

On January 31, 1986, JSC employees, contractors, and invited guests gathered to pay their respects to the crew of STS-51L at an outdoor memorial service. Thousands of mourners attended the event, located in an open area surrounded by several of the center's historic buildings. Pres. Ronald Reagan came to comfort the families, the workforce, and the nation. His eulogy remembered the seven astronauts who died as a result of a faulty O-ring design and promised the families, NASA, and the country that space exploration would continue. Two and a half years later, the STS-26 mission fulfilled the president's vow by returning the Space Shuttle to flight. During the flight, the crew paid homage to the fallen astronauts. "Dear friends, we have resumed the journey that we promised to continue for you. Dear friends, your loss has meant that we could confidently begin anew. Dear friends, your spirit and your dream are still alive in our hearts."

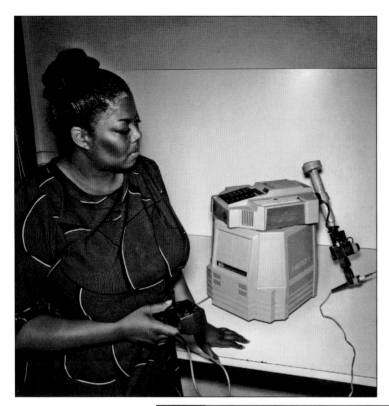

In the mid-1980s, the Avionics Division built the HERO-1 robot, representing one of JSC's earliest experiments in robotics. Center leadership recognized that the discipline would be essential to future spaceflight programs and in 1990 established the Robotics Division in the Engineering Directorate. Engineers envisioned that robots would one day assist flight crews on orbit.

Researchers examine a computer-controlled two-axis positioner, part of the Synthetic Vision System, developed by JSC's Tracking and Techniques Branch. Funded by the Technology Utilization Office and the US Army, the system—once perfected—would allow space cameras to identify and search for objects in space, like malfunctioning satellites, which could then be brought to a future space station for repair.

Aaron Cohen began working at the Center in 1962 as an engineer. He later moved into spacecraft management and from 1972 to 1982 directed the Orbiter Project Office. As JSC's fifth Center Director (1986–1993), Cohen oversaw the Return to Flight effort after the *Challenger* accident and helped establish the new visitor's center, Space Center Houston.

Center employees passed the Olympic flame as part of a torch relay in the Clear Lake area. Houston hosted the US Olympic Festival in 1986, and the torch that 35 employees carried across the site was eventually brought to Houston's Astrodome for the official opening ceremonies and the lighting of the Olympic torch.

In 1987, JSC's Structures and Mechanics Division tested a telerobotic device that could be used to assemble structures in outer space. The operator shown here used a hand controller to maneuver the simulated Remote Manipulator System, or robotic arm, that astronauts would use to build the future space station.

Looking for a way to simulate microgravity on Earth, biotechnology researchers in the Medical Sciences Division made extraordinary medical contributions through pioneering efforts with rotating wall bioreactors. Culturing delicate living cells in a slowly rotating cylinder resulted in lower-stress environments and natural, three-dimensional tissue growth. The devices, later flown into space, helped develop pharmaceutical products for fighting cancer and other diseases.

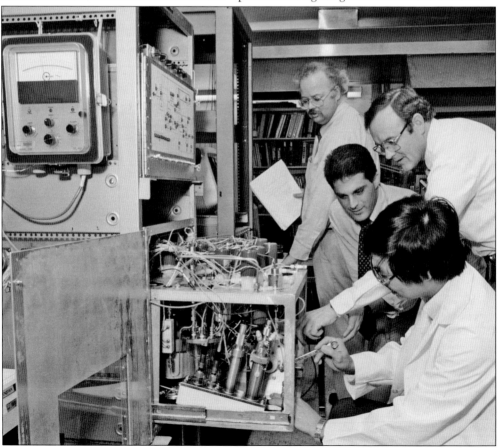

A wide-angle photograph of the Weightless Environment Training Facility captures the activities and people involved in JSC's underwater training program. Scuba divers, suit technicians, medical doctors, safety engineers, and test conductors participated in "test runs" in the facility. Consoles ran along the perimeter of the pool, while submersible cameras allowed trainers to monitor underwater operations. An outline of the payload bay can be seen below the surface of the water.

A diver in the Weightless Environment Training Facility monitors a suited astronaut as the crewmember works with a mock-up during a test of the Space Station *Freedom* airlock concept. This training session also evaluated techniques for manually handling the Orbital Refueling Unit, which spacewalkers had tested in orbit in 1984. JSC built, designed, and tested the Orbital Refueling System, capable of refueling satellites in space.

Multiple flight control rooms are the activity focal points in the Mission Control Center in Houston, where controllers have supervised spaceflight missions since Gemini 4. Located in Building 30, Mission Operations Control Room (MOCR) 1 directed the Apollo 7 flight, the Skylab manned missions, the Apollo-Soyuz Test Project, and unclassified Shuttle flights through STS-76 in 1996. After the four initial Shuttle Program test flights, the MOCRs were designated as Flight Control Rooms (FCRs), and MOCR-1 became FCR-1 (seen above during STS-29). In 1998, after a major remodeling, FCR-1 became the Life Sciences Center for International Space Station (ISS) payload control operations. Eight years later and after another technological overhaul in 2006, ISS flight controllers moved into the totally revamped FCR-1, still in use as of 2013.

Four

1990s
FORGING PARTNERSHIPS

During the 1990s, the Johnson Space Center proved to be instrumental in laying the foundation for future space exploration. After the Cold War ended, America's space adversary became its partner and JSC built on the collaboration established in the 1970s with the Apollo-Soyuz Test Project. The center led the US and Russian program known as Shuttle-*Mir*, where from 1995 to 1998, a series of nine Shuttle missions ferried seven American astronauts to the Russian *Mir* Space Station for long-duration stays. This joint effort helped to cement the relationship between the two space programs and served as a critical step towards the future International Space Station (ISS).

In 1997, the Neutral Buoyancy Laboratory (NBL) opened and served an important role in training the astronauts who built the ISS. The 6.2-million-gallon pool allowed spacewalkers to practice the tasks they would complete on orbit by simulating the zero-gravity conditions of space. Training in the NBL and its predecessor, the Weightless Environment Training Facility (WETF), was also critical to the three crews that serviced and maintained the Hubble Space Telescope in the 1990s.

In addition to Hubble, the Shuttle deployed two other Great Observatories for astronomy in the 1990s: the Compton Gamma Ray Observatory and the Chandra X-Ray Observatory. The 1999 Chandra mission was also notable for being the first Shuttle mission to have a woman as its commander. Only a few months earlier, one of the original astronauts, John Glenn, 77, returned to space aboard *Discovery*, and once again the public's interest soared as it had more than 30 years earlier, when he orbited the Earth in *Friendship 7*.

Thousands of visitors travel to Houston to get a first-hand look at Mission Control and the astronaut training facilities. To accommodate the increasing numbers, the small onsite visitor's center was replaced in 1992 with Space Center Houston, a spacious facility adjacent to JSC.

And in November 1998, the crew of STS-88 assembled the first core modules of the ISS that would serve as an orbiting laboratory for multinational crews and space explorers, beginning the next chapter in human spaceflight.

Gov. Ann Richards of Texas (right) and Sen. Barbara Mikulski of Maryland tried out the mission specialist and pilot seats in the Shuttle Flight Deck Full Fuselage mock-up in the Building 5 Training Facility during a 1991 visit. Political figures and dignitaries often visit the center to gain first-hand knowledge in support of the space program.

Queen Elizabeth II of England and Prince Philip visited Johnson Space Center in 1991. They enjoyed a luncheon in Building 9 and a tour of the Mission Control Center. As the queen greeted the crowds in front of Building 30, an employee presented her with a bouquet of yellow roses.

Lunar Test Article 8 (LTA-8) was moved to Space Center Houston via crane prior to the grand opening of the new visitor's center in October 1992. As the first man-rated Lunar Module, the ascent and landing stage trainer was used by Apollo astronauts during preparation for the historic Moon landings. Tested in Chamber B of the Space Environment Simulation Laboratory, LTA-8 demonstrated that the Lunar Modules' environmental control system could provide a habitable environment and temperature for the crew under simulated lunar conditions. However, since the module was designed for use in one-sixth of Earth's gravity, it was not feasible to keep the LTA-8 manned for the entire duration of testing. The crew occupied the Lunar Module in shifts ranging from 9 to 12 hours, ingressing and egressing the vehicle under full-vacuum conditions. LTA-8 continues to be a featured attraction for visitors.

Space Center Houston (white building, bottom center) opened its doors to the public in October 1992. Years of planning and preparation went into the NASA "experience center," developed by the nonprofit Manned Space Flight Education Foundation, Inc., and designed by Walt Disney Imagineering. Covering 50 acres of land adjacent to JSC, the 183,000-square-foot facility was designed to give visitors an understanding of the complexity of spaceflight and a behind-the-scenes look at JSC. Using theatrical and hands-on presentation techniques, the center allows visitors to experience some of the challenges of space travel. Historical artifacts and hardware, such as spacesuits, space vehicles, and a Moon rock, are on display. A tram takes visitors on a tour of training facilities and Mission Control. With an emphasis on education, Space Center Houston hosts teacher workshops and conferences, school programs, camps, and outreach programs.

As technological needs changed, Building 30's Mission Control Center (MCC) grew to include a five-story addition, Building 30 South. Housed on the second floor, the White Flight Control Room (FCR) began operation in 1995 with a ribbon cutting ceremony officiated by Center Director Carolyn Huntoon (above). First used for the orbit phase of STS-70, this facility became the primary Space Shuttle Control Room with the first dedicated complete flight, STS-77, in May 1996. The White FCR computer consoles provided status and event displays, allowing flight controllers to monitor and control each mission as well as network MCC resources.

The Wake Shield Facility (WSF) was designed and built by the Center for Advanced Materials at the University of Houston. Seen here in Building 49 before testing, the WSF was deployed in the wake of the Space Shuttle and redirected atmospheric particles around the sides, leaving an ultra-vacuum in its wake that could be used for materials processing.

John Glenn (seated in cockpit) was the first American astronaut to orbit Earth. After a long career as a US senator, he returned to NASA to fly on STS-95 and, at age 77, became the oldest person to go into space. The flight allowed researchers to gather information on weightlessness and other aspects of spaceflight from the same person on two separate missions 36 years apart.

Engineers developed the Simplified Aid for EVA Rescue, or SAFER, for emergency use during extravehicular activities (EVAs), or spacewalks. The SAFER, a small, self-propelled backpack, allows astronauts to maneuver themselves back to safety in the event of a tether break and was first tested on STS-64 in 1994. Astronauts trained to use the SAFER on the air-bearing floor in the Shuttle Mock-up and Integration Laboratory in Building 9.

For long-duration spaceflight, crewmembers were prepared for any emergency, including medical problems. The Shuttle-*Mir* program, Phase 1 of the International Space Station, created training opportunities for American and Russian crews to practice procedures while still on Earth. JSC medical doctors trained the international teams on detailed techniques, including maintaining a patient's airway.

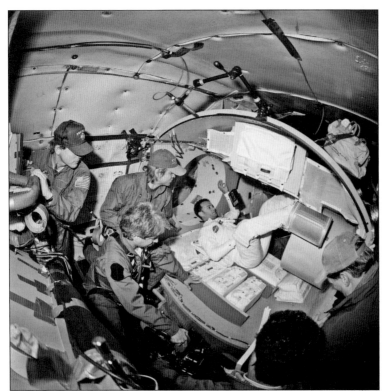

Director Ron Howard, actor Tom Hanks, and the *Apollo 13* film crew worked aboard the KC-135 aircraft during the filming of the movie's zero-gravity scenes in 1995. The story followed crewmembers and the Mission Control teams as they worked through the life-threatening accident and problems encountered during the ill-fated Apollo 13 mission.

The Sonny Carter Neutral Buoyancy Laboratory accommodates spacewalk preparation and training in a weightless environment, critical to the success of the International Space Station program. In 1998, astronauts trained for three spacewalks scheduled during the first ISS assembly mission, which included attaching power and data cables between full-scale mock-ups.

The Full Fuselage Trainer in Building 9 allowed STS-61 astronauts to practice the first Hubble Space Telescope (HST) servicing mission using an inflatable mock-up. Deployed in 1990, a flaw in the mirror prevented the HST from sending clear images back to Earth. The repair mission required five spacewalks for the crew to complete the necessary repairs.

The Space Shuttle Emergency Slide provided the Orbiter flight crews with a rapid and safe emergency egress through the Orbiter middeck side hatch after a normal opening or an emergency landing. The crews practiced procedures in the Crew Compartment Trainer in Building 9 while wearing their launch and entry suits.

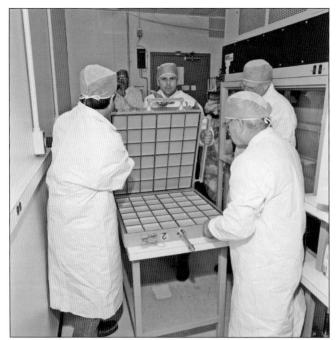

The *Mir* Environmental Effects Payload, deployed on the *Mir* Space Station and retrieved after 18 months in space, studied the frequency and effects of both human-made and natural space debris. JSC's Orbital Debris Collector experiment captured residues of the particulate environment by exposing two trays with highly porous aerogel as the collector medium.

As an international partner, Japan played a significant role in the development and building of the International Space Station. During a visit to the Johnson Space Center in 1997, the president of the Japanese Space Agency toured the facilities, including a review of flight simulation equipment in Building 5.

Pres. Bill Clinton visited the Johnson Space Center in 1998 and toured the Shuttle and International Space Station training facilities. While at the center, he was briefed on various NASA programs and met with employees, agency officials, and former US Senator John Glenn, who was training for his upcoming STS-95 mission.

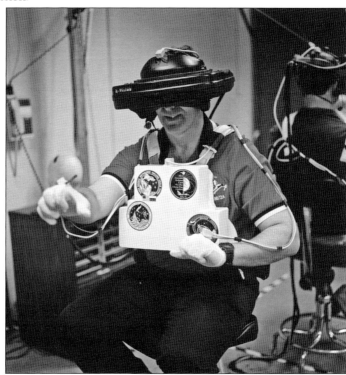

Virtual reality was one of the training tools used to prepare astronauts to assemble the International Space Station. With a helmet and special gloves, the astronauts could preview exactly what they would expect to see, simulate actual movements around the hardware, and practice techniques and procedures in the computer-generated environment.

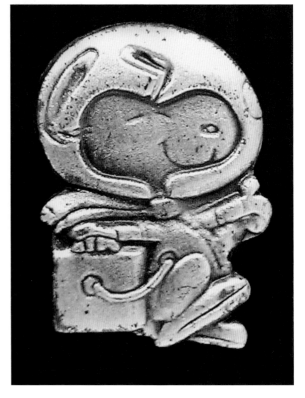

Charles Schulz, creator of the *Peanuts* comic strip, and his wife, Jean, sit in the mission specialist seats on the Shuttle Flight Deck Full Fuselage mock-up in 1998. In the 1960s, NASA introduced an award featuring the well-known Snoopy character dressed as an astronaut to be given by astronauts to employees in recognition of their outstanding contributions. Schulz designed the image for this pin, known as the "Silver Snoopy" (right), which was first presented in 1968. Award ceremonies continue to this day for "professionalism, dedication and outstanding support that greatly enhanced spaceflight safety and mission success." (Snoopy image courtesy of Peanuts Worldwide, LLC.)

STS-93, the first Space Shuttle mission commanded by a woman, Eileen Collins, left its mark on history and in the sky over JSC's Rocket Park during reentry on July 27, 1999. It was only the 12th night landing for the Shuttle program, and observers across the flight path were treated to a light show as *Columbia* streaked toward Kennedy Space Center in Florida for a safe landing. The primary payload for this mission was the Chandra X-Ray Observatory, the third of NASA's four Great Observatories. Successfully deployed during STS-93, Chandra detects and images X-ray sources that lie within this solar system and those billions of light years away. The results from Chandra help provide insights into the universe's structure and evolution.

A test section of the International Space Station truss arrived at Building 49 for vibration-acoustic testing simulating Shuttle launch conditions. The first of 11 segments connected in orbit to build the backbone of ISS, the S0 truss segment was the junction from which external utilities were routed to the pressurized modules.

The Transhab module, an expandable habitat concept for use as living quarters for astronauts, was tested in Chamber A of the Space Environmental Simulation Laboratory (SESL). The vacuum chamber simulates the atmospheric pressure and temperature ranges vehicles and components may encounter in space. Since 1966, hardware from every major human spaceflight program has been tested in one of the two SESL chambers.

The Space Vehicle Mock-up Facility in Building 9 (pictured above in 1999) contains facilities to support astronaut training and engineering activities. Building 9 housed the Technical Services Division in the 1960s, and the first addition, 9A, was added in 1974 to house the Shuttle mock-ups. In 1988, construction was completed on the next addition, 9B, to support the International Space Station. The Space Shuttle Orbiter (background, top of photograph) and ISS (foreground) trainers were built to achieve as much realism as possible during practice sessions and flight simulations. Flight crewmembers, controllers, scientists, and instructors experienced high-fidelity work and living environments in these mock-ups. Ongoing ISS training on the station's modules, nodes, and truss structure provided practice opportunities for the international crews. In 1985, the STS-61B crew was the first to train for ISS assembly with experiments designed to study how quickly astronauts could adapt to constructing space structures during extravehicular activities. Today, training continues to support ISS operations and to ensure maximum safety and mission success on orbit.

Over the years, NASA has participated with the Houston Livestock Show and Rodeo in educational and community outreach projects. In 1997, JSC officials introduced dignitaries from the Russian Space Agency to a Texas welcome and fireworks display at the event held in the Astrodome. For 46 years, the events were showcased at the Astrodome complex.

JSC engineers and scientists have a long tradition of outreach with area schools and universities. Educational opportunities include school visits, such as the one above during National Engineers Week, where students can reach out and touch groundbreaking NASA technologies, participate in experiments, and learn about careers in the space program.

The Longhorn Project at JSC began in 1997 with the release of two longhorns on land that was once home to an entire herd. Local high school students are provided with a unique hands-on learning opportunity for study and research while celebrating a uniquely Texan heritage. The Longhorn Project is supported by JSC, Clear Creek Independent School District, Houston Livestock Show and Rodeo, and Texas Longhorn Breeders Association of America. Center Director George Abbey (above, right) was instrumental in bringing the longhorns to the property and participated in the dedication ceremony of the limestone Western Heritage Pavilion.

The Advanced Life Support Program conducted three manned experiments in sealed chambers to test the efficiency of regenerative plant-based systems. During Phase 1 of the program, one test subject (above) lived in the 24-foot-long chamber in Building 7 for 15 days. Divided into two sections, the enclosed environment held a plant growth compartment and an airlock for living quarters. The wheat chosen for this phase exceeded the amount of oxygen necessary to support one person for the duration of the trial. For humans to travel for long periods in space, effective ecological life support systems must be developed to produce food, recycle water and solid materials, and regenerate oxygen. Carbon dioxide produced by humans is used to grow plants that in turn provide oxygen and food. Plants would also provide a much-needed diversion and psychological boost to a long-term stay in a hostile environment or on another planet.

Five

2000s
CONTINUAL PRESENCE IN SPACE

In the fall of 2000, Expedition 1, a three-man crew consisting of an American astronaut and two Russian cosmonauts, began humanity's continued presence in space aboard the International Space Station (ISS). JSC, home to the ISS Program Office, coordinated with 15 nations on this global project, which required space agencies, engineers, scientists, astronauts, and cosmonauts to work together to accomplish this endeavor. A multinational effort, crews trained not only in Houston but also at facilities across the world. Today, the station is the largest object, other than the Moon, to orbit the Earth.

The Space Shuttle played a central role in delivering ISS modules, hardware, supplies, and flight crews to orbit until Saturday morning, February 1, 2003, when *Columbia* was lost over East Texas and Louisiana upon reentry. The heat shield of the Orbiter's left wing had been irreparably damaged during launch after being hit by a piece of foam that had broken loose from the external tank. All flights were put on hold until the agency upgraded specific Shuttle systems and instituted changes to NASA's culture. In the summer of 2005, two and a half years after the accident, *Discovery* flew its second successful Return to Flight mission. In the interim, Russian Soyuz capsules launched from the Baikonur Cosmodrome in Kazakhstan transported the Expedition crews to the station.

In the midst of the flight hiatus, on January 14, 2004, Pres. George W. Bush announced a new vision for space exploration calling for a return to the Moon and travel to Mars and other destinations. The ISS was scheduled for completion by 2010, after which NASA would retire its 30-year-old workhorses in the Shuttle fleet. NASA designated JSC as the leader of the new Constellation program implemented to achieve the president's vision. JSC, which engineered spacecraft, began to design and develop the Crew Exploration Vehicle, known as the Orion spacecraft, the first new vehicle in more than three decades, as well as the Altair lunar lander and a new spacesuit system.

The Expedition 1 crew, the first International Space Station crew, posed for photographs prior to their launch aboard a Soyuz from the Baikonur Cosmodrome in Kazakhstan (left). Their residency on the ISS began on November 2, 2000, and established an uninterrupted human presence in space that continues today. The American commander and two Russian cosmonauts activated systems, unpacked equipment, and were visited by three Space Shuttle crews. Two unmanned Russian Progress vehicles delivered supplies to the space travelers. In March 2001, STS-102 returned the crew safely to Earth after ferrying the Expedition 2 crew to the station. Below, prior to launch, the Expedition 1 crew trained at JSC with mock-ups of both the Soyuz and the Shuttle in Building 9.

During a fly-around aboard the Soyuz in February 2001, the Expedition 1 crew photographed the earliest phases of the ISS configuration. Assembly of the International Space Station began in November 1998 with the launch of the first segment, the Russian Zarya (meaning "dawn" in Russian). Two weeks later, *Endeavour* launched with the STS-88 crew and America's Unity module in the Shuttle's payload bay. Using *Endeavour*'s robot arm, the crew docked the two components and conducted two spacewalks to connect power and data cables. In July 2000, the Zvezda service module launched from Baikonur and docked with the ISS to become the third component. The STS-106 crew made the necessary connections to Zvezda in September of that year, after which the module served as living quarters with life support and communication systems, electrical power distribution and data processing, flight control, and propulsion. Two more Shuttle flights added segments of the Integrated Truss Structure, the backbone of the station.

The Blue Flight Control Room (FCR) was designated the control room for International Space Station flights from 1998 until 2006. Located adjacent to the White FCR in Building 30 South of the Mission Control Center, this was originally the Special Vehicles Operations Room, used for monitoring spacewalks and other specialized activities. New systems had to be developed and installed for ISS activities, including telemetry and station commands. In 1998, the first station assembly flights launched, and the Blue FCR began operating around the clock to support the operations. From this room, flight controllers monitored the ISS Expedition missions that began in Kazakhstan with the crews launching on a Russian Soyuz spacecraft (seen on monitor screen). Traditionally, as with all the control rooms from Gemini through ISS, after the completion of a successful mission, flight control personnel participated in a special ceremony to hang the mission patch on the control room walls (seen at upper right).

Astronauts discuss data at the CapCom (short for Capsule Communicator) console in Mission Control. A position created during the Mercury program, the CapCom relays information between spaceflight crews and Mission Control. Until 2001, the position was reserved for astronauts only, but additional personnel were trained to meet the needs of long-duration missions for the ISS that require crew activities to be monitored from Mission Control 24 hours a day.

Engineers in Mission Control inspected the Shuttle's heat-protecting tiles for damage using images captured by the Orbiter Boom Sensor System (OBSS), first flown in 2005 on STS-114, the Return to Flight mission after *Columbia*. Using the robotic Canadarm, astronauts maneuvered the OBSS instrument with its package of cameras and lasers around the Shuttle's underbelly and then sent the pictures back to Earth for review.

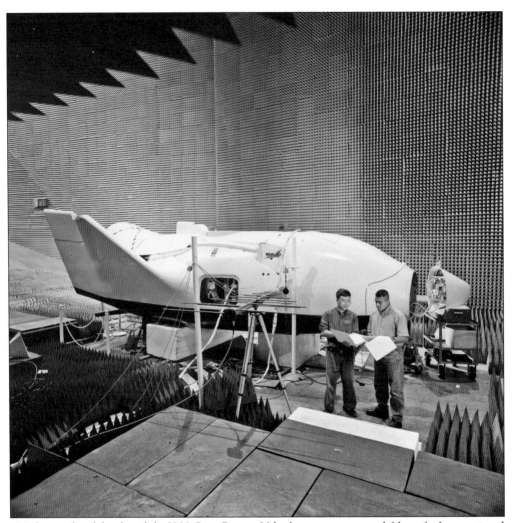

JSC designed and developed the X-38 Crew Return Vehicle as an emergency lifeboat for International Space Station crews and as an alternative to the Russian Soyuz transport with its maximum carrying capacity of three people. Engineers planned to use the Space Shuttle to ferry the X-38 to the ISS, where the spacecraft could be docked for up to two years. The project aimed to use commercial off-the-shelf technology for up to 80 percent of the vehicle, allowing engineers to minimize costs and use the savings to build more vehicles. Above, during its development phase, a prototype of the vehicle was tested in the JSC anechoic, or echo-free, chamber in Building 14 that engineers had built in the 1960s to test Apollo spacecraft. The walls of the facility are completely covered in microwave material that absorbs electromagnetic energy to simulate the environment of space. The test model of the X-38 was housed in Building 200, then later in Building 10. The X-38 program was canceled in 2002.

The JSC family shared their grief and support after the tragedy of September 11, 2001, by signing large banners posted at sites along the center's perimeter. Acting Center Director Roy Estess (above) guided the center through those first uncertain days after the attacks. Estess, a former test engineer during the Apollo program, also served as director of the Stennis Space Center in Mississippi.

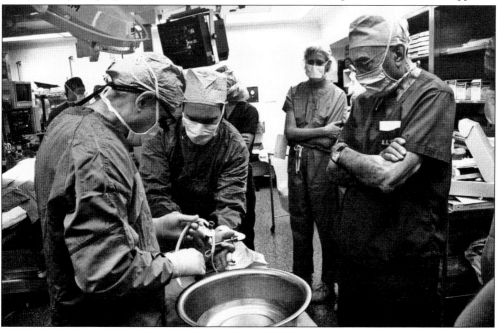

The MicroMed DeBakey VAD (ventricular assist device) is one example of how NASA's innovations in science and engineering impact the rest of the world. NASA engineers first began working with renowned heart surgeon Dr. Michael DeBakey (pictured at right at Texas Medical Center) in the mid-1980s on the revolutionary heart pump. The device received US Food and Drug Administration approval in February 2004 for use in critically ill children awaiting heart transplants.

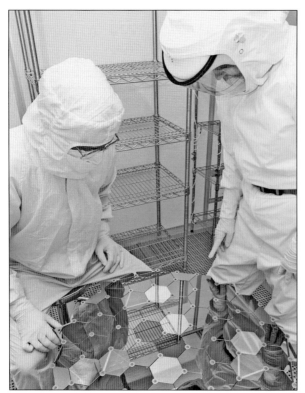

The Genesis mission was launched in August 2001 to collect samples of solar wind particles and return them to Earth for analysis. Prior to launch, technicians inspected one of five translucent arrays in JSC's cleanroom, where the arrays were assembled from individual collectors made of a variety of ultrapure materials. Each delicate wafer was handled with special instruments created solely for the purpose of keeping the arrays ultraclean and free from contamination (below). The samples returned to Earth in 2004, but the capsule's parachutes did not deploy, and it slammed into the ground. The team at the Genesis curation laboratory devoted extensive effort to identifying, cleaning, and cataloging the thousands of collector fragments produced by the crash. Their efforts have revealed information about processes and conditions in the early history of the solar system.

The STS-107 mission ended tragically on February 1, 2003, just 16 minutes from home. During *Columbia*'s January 16 launch, a piece of foam from the external tank struck the leading edge of the left wing of the Orbiter. At reentry, that breach in the thermal protection system allowed heat to build up and caused the Shuttle to lose control and break apart over Texas. Four days after the terrible loss, thousands of Johnson Space Center employees and invited guests joined with the crewmembers' families to remember and pay tribute to the STS-107 crew. In an emotional event, Pres. George W. Bush addressed the crowd with words of comfort and honored *Columbia*'s last crew: "We remember not only one moment of tragedy, but seven lives of great purpose and achievement. For these seven, it was a dream fulfilled. Each of these astronauts had the daring and discipline required of their calling. Each of them knew that great endeavors are inseparable from great risk. And each of them accepted those risks willingly, even joyfully, in the cause of discovery."

STS-114, the first Shuttle mission after the *Columbia* accident, launched on July 26, 2005, and delivered needed supplies to the International Space Station. The major focus of the mission was to test and evaluate new flight safety techniques for the thermal protection system in-flight inspection and repair. Prior to the launch, the flight director team explained the mission goals at a press briefing in the Teague Auditorium, JSC Building 2.

After the successful Return to Flight mission, the STS-114 crew and Center Director Jefferson Howell (left) took part in pregame ceremonies during NASA Night with the Houston Astros. NASA and the baseball team, named in 1965 because of Houston's central role in the space program, have a long-standing relationship. To honor the *Columbia* crew, the Astros wore the STS-107 patch on their uniforms in 2003.

The Virtual Reality Laboratory in Building 9 provides immersive simulation training for astronauts as they prepare for spacewalks, construction and maintenance, and robotic manipulation tasks as International Space Station crewmembers. The real-time graphics and motion simulators allow astronauts to practice planned activities while on Earth. The Expedition 11 crew (above) trained on a computer interface with virtual reality training hardware and software that prepared them for their upcoming mission.

Orbital debris, both man-made and naturally occurring, can cause damage or catastrophic failure to a spacecraft. The White Sands Test Facility (WSTF) Remote Hypervelocity Test Laboratory in New Mexico simulates the effect of these impacts on spacecraft, satellites, and spacesuits. The lab contributed to the development of methods to better protect vehicles and humans in the dangerous environment of space. WSTF is managed by the Johnson Space Center.

At the beginning of the space program, the need to test spacecraft against the harsh conditions encountered in space became evident. In 1963, work began on the Space Environmental Simulation Laboratory, and after the two vacuum chambers were built, Building 32 was constructed around the chambers to enclose and house the laboratory. The chambers simulate the vacuum and temperatures encountered more than 130 miles above Earth, plus have the added advantage of returning the spacecraft and crew to the Earth's atmospheric conditions in a matter of seconds. The 40-foot-tall, 40-ton door to Chamber A (above) is closed with the aid of hydraulic hinges and can be moved easily by two people. Once inside the stainless steel tank, components are placed on the rotating floor for pressurized thermal testing. During the early spaceflight programs, entire missions were simulated within the chamber, including test subjects or astronauts living and working in their capsules. Chamber B, a smaller chamber built for testing spacesuits and hardware, allows astronauts to practice deployment of equipment in space, practice repairs, and certify tools.

The Space Environmental Simulation Laboratory chambers, normally used to analyze hardware during a simulated trip to space, occasionally are used for more earthly endeavors. After a water pipe burst at a local university library in 2006, approximately 1,800 drenched books were placed in Chamber B to allow the water to sublimate or vaporize. With a few vacuum cycles, the books were dry and ready to be returned to the library.

The Space Station Airlock Test Article (SSATA) is located in Building 7 in the Crew Systems Laboratory and used to support the International Space Station program. The man-rated vacuum chamber allows testing, verification, certification, and crew training with airlock and extravehicular activity (EVA) hardware. The SSATA provides flight-like simulation of airlock and EVA operations in pressures ranging from vacuum to one atmosphere.

In 1977, the Saturn V rocket arrived at JSC to become an open-air exhibit near the center's main entrance. After 30 years on display in Houston's hot and humid climate, the artifact required extensive restoration and a building to protect the national treasure from the elements. A grand opening ceremony of the new facility was held on July 20, 2007, the 38th anniversary of the Apollo 11 mission.

The Ballunar Liftoff Festival began from a simple idea to commemorate the anniversary of the first manned balloon flight in France 210 years earlier. Held at the Johnson Space Center, the annual event has continued since the first celebration in 1993. One of the largest national navigational and maneuvering competitions for hot-air balloons, the three-day event features colorful and visually stunning balloons from all across the country.

Crew debriefing and awards presentations after Space Shuttle flights and Expedition missions were held at Space Center Houston in the largest giant-screen theater in Texas. Five stories tall and equipped with the latest in digital projection technology, the venue provided the perfect backdrop for the imagery presented by the crews for their families and coworkers. Seen on the stage at the bottom of this photograph is the STS-118 crew.

Astronaut and teacher Barbara Morgan spoke with students at Space Center Houston as part of NASA's educational outreach program. In 1985, Morgan qualified as backup for the first Teacher in Space program; in 1998, she became an astronaut. She was a mission specialist in 2007 for STS-118, which carried into orbit an education payload of 10 million basil seeds. The seeds were later used in classrooms throughout the nation.

During a visit to JSC in 2007 with her husband, former First Lady Barbara Bush spoke with Center Director Michael Coats and his wife, Diane. Coats began his career with NASA as one of the 35 members of the groundbreaking 1978 astronaut class, the first to include women and ethnic minorities. A veteran of three spaceflights, Coats served as director from 2005 to 2012.

The Johnson Space Center High Altitude Research Program operates the only two WB-57 aircraft still in use today. Seen here above the nearby San Jacinto Monument, the mid-wing, long-range aircraft are capable of operation for extended periods of time at altitudes in excess of 60,000 feet. By providing a unique airborne platform, the program supports scientific research and advanced technology development for government agencies, academic institutions, and commercial customers.

While mounted on the Shuttle Carrier Aircraft (SCA), the Space Shuttle *Endeavour* flew over JSC in December 2008. The 1,625-acre space center can be seen in the background of the photograph, taken from the rear station of a NASA T-38 aircraft flying over Clear Lake. The extremely clear weather set the stage for a beautiful fly-over event and an outstanding photo opportunity as the SCA and *Endeavour* slowly circled over the area. Employees, schoolchildren, and residents stood outside to witness this rare event. The SCA, a modified Boeing 747, was used to ferry Space Shuttles from landing sites back to the Shuttle Landing Facility at the Kennedy Space Center in Florida for processing and preparation for the next mission. The Orbiters were placed on top of the SCAs by mate-demate devices, large gantry-like structures that hoisted the Orbiters off the ground for post-flight servicing and then mated them with the SCAs for ferry flights.

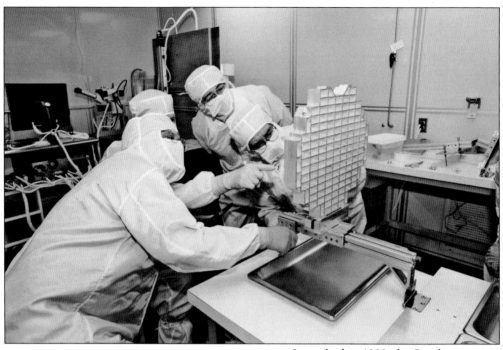

Launched in 1999, the Stardust mission was the first to return particle samples from a comet and interstellar space. In January 2004, as the spacecraft encountered the Jupiter Family comet Wild-2 and as it passed through the coma (area around the nucleus of the comet), a tray of silica aerogel blocks was deployed to capture several thousand coma dust grains. Another collector tray was exposed to the interstellar dust stream for 196 days during two periods prior to the comet encounter. In January 2006, the sample return capsule safely parachuted back to Earth and was flown to the Stardust laboratory at Johnson Space Center. Following the preliminary examination for each sample, materials from the interstellar tray were made available to the international research community for detailed analyses. Since the return, over 1,200 comet samples have been allocated for research.

World-renowned theoretical physicist Stephen Hawking visited the full-scale mock-ups in Building 9 in 2007. Hawking, best known for his scientific writings and lectures on the origins of the universe, has an increasing disability due to a degenerative nerve disease known as amyotrophic lateral sclerosis. Wheelchair bound on Earth, he also traveled to Florida's Kennedy Space Center to take a ride on a zero-gravity aircraft and experience weightlessness.

The Hubble Space Telescope (HST), the first of NASA's four Great Observatories, was deployed in 1990 and required occasional upgrades by Shuttle crews. Astronauts trained extensively at JSC to prepare for these tasks, including the final HST mission, STS-125. Five spacewalks were conducted to extend the life of the telescope after the HST was grappled by the robotic arm (seen through the Shuttle window) then docked in the Orbiter's payload bay.

STS-118 was the 22nd Shuttle flight to ISS and brought a third starboard truss segment to the growing outpost. Before the flight, astronauts practiced adding the two-ton, 11-foot-long spacer to the truss in the Sonny Carter Training Facility Neutral Buoyancy Laboratory. Built to accommodate two separate training activities simultaneously, the NBL provides controlled neutral buoyancy operations simulating the weightless environment encountered on spacewalks or extravehicular activities.

As the Space Shuttle program and construction of the International Space Station came to an end, NASA opened the NBL to use by commercial partners. The first company to take advantage of the opportunity brought its autonomous underwater vehicle *Endurance*, a prototype for developing and testing capabilities that would be needed to explore Europa, Jupiter's sixth moon. The event was monitored by the company and NBL staff.

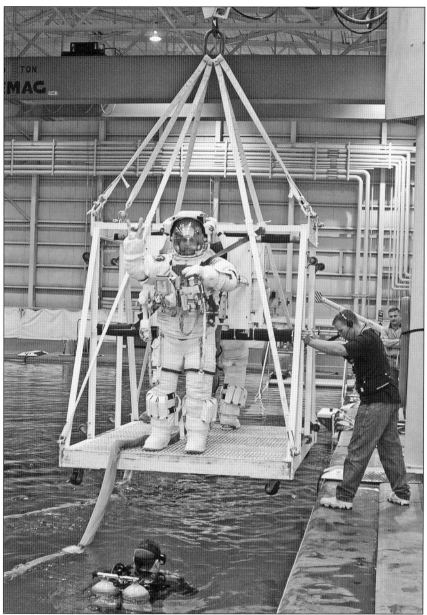

Practicing extravehicular activities (EVA) in a weightless environment was paramount to the building of the International Space Station. The Neutral Buoyancy Laboratory pool at the Sonny Carter Training Facility allows crews to train using full-scale mock-ups of the ISS modules and components, as well as the EVA spacesuits necessary for protecting them from the harsh environment of space. Scuba divers aid the astronauts and monitor their activities throughout each practice session in the 6.2-million-gallon pool. At 202 feet in length, 102 feet wide, and 40 feet deep, the NBL can accommodate two separate training activities at the same time. Technicians in the control areas provide resources and expertise for all disciplines involved in the execution of spacewalk training, including communications, safety, and suit specialists. A complete medical team monitors the condition of all dive personnel, and a fully configured hyperbaric chamber located nearby can be used for emergency decompression sickness if needed.

Humans who explore other planets or moons will require a surface transport vehicle capable of carrying astronauts and their equipment. Riding instead of walking would allow crewmembers to conserve important spacesuit consumables and allow spacewalk activities at a higher performance level and for longer periods of time. The Science Crew Operations and Utility Testbed (SCOUT), above, was designed to develop and test advanced rover technologies and operations concepts.

The Johnson Space Center took a step toward renewable energy by constructing the Multi-Platform Renewable Energy System at the Aaron Cohen Child Care Center. The goals of the project were to gain experience with renewable energy technologies, to collect real field data to confirm the viability of these systems, and to gain an understanding of large, surface-based photovoltaic arrays that would be necessary for exploration of other planets or moons.

The Pyrotechnics Test Facility in Building 352 has a long history of providing safe, reliable testing of the explosives used to launch spacecraft. Engineers used the cut-away test article seen here to study a possible ignition system for future launch vehicles. Building 352 houses an explosives loading and handling room and pyrotechnics storage in earth-covered bunkers.

In March 2007, a Mark III advanced spacesuit technology demonstrator participated in a systems engineering and integration test in the Space Vehicle Mock-up Facility at the Johnson Space Center. Data collected in the test is being used by JSC engineers as they continue to develop the next generation of spacesuits for future space exploration.

In 2004, Pres. George W. Bush announced his Vision for Space Exploration, a plan to return humans to the Moon and continue onward to Mars. In response, Johnson Space Center began designing the Crew Exploration Vehicle (CEV), along with the Ares rockets that would boost the capsule to space. Assembly of the full-size mock-up of the CEV began in 2005 (above). During development, test subjects wearing launch and entry suits helped engineers analyze the strengths and weaknesses of the CEV's proposed seat layout. The vehicle, which resembles the Apollo Command Modules of the 1960s, is three times larger and was designed to support up to six crewmembers on future missions.

Six

2010s
Halfway to Everywhere

As the Johnson Space Center entered its 50th year, the center began a time of transition. Just the year before, JSC was responsible for the management of three of NASA's human spaceflight programs: Space Shuttle, International Space Station (ISS), and Constellation—the program to fly astronauts to the Moon and destinations beyond. However, Constellation was canceled in February 2010, and in July 2011, the last flight of *Atlantis* ended the 30-year Shuttle program. Only the ISS remained, with JSC directing the operations for the orbiting space laboratory.

During this transition, center leaders worked hard to foster partnerships with industry and academic institutions to spark innovative ways to sustain and maximize JSC's core competencies.

As the center's next 50 years of history begin, JSC's leadership in human spaceflight remains central to the exploration of the solar system. Despite the hardships of sustained budget cuts, JSC remains a leader in fields such as robotics and propulsion technology to allow astronauts to travel farther than ever before. Engineers' work on the Orion capsule, also known as the Multi-Purpose Crew Vehicle, helps commercial space transportation companies develop the spacecraft needed to deliver supplies to the ISS and provide safe and routine space travel for flight crews.

Based upon the knowledge gained through the years, JSC provides additional innovative development, laying the groundwork for a sustainable program of exploration. The center's pioneering spirit, established by the first generation of space explorers, remains essential to accomplishing NASA's current goal of landing on an asteroid by 2025 and reaching Mars by the mid-2030s. And for the next decade, JSC will direct the ISS as the multinational crews engage in cutting-edge research in areas that could prove crucial in future missions in addition to helping improve life here on Earth.

More than 50 years after its establishment, JSC continues to provide the technical know-how and unfailing dedication necessary to explore the next frontier and shape the future of human spaceflight.

The STS-135 crew practiced rendezvous and docking with the International Space Station in the Systems Engineering Simulator in Building 16. The reconfigurable cockpit mock-up, located within a 24-foot-diameter hemispherical dome, projects one continuous image onto its walls. This "real-time, crew-in-the-loop" engineering simulator models an external environment that can be viewed through the windows, allowing crewmembers and engineers to perform tests, evaluations, and training in a controlled environment. Although STS-135 was the last mission of the Space Shuttle program, the simulator continues to support both crew training and engineering analysis of on-orbit operations for ISS. It is also a crucial tool in the development of advanced programs where the interface between humans and machines, or between vehicles and subsystems, is important. Throughout the history of NASA's human spaceflight programs, simulations conducted at the Johnson Space Center have served the important function of providing the means to train crews and evaluate spacecraft in a realistic environment.

Astronauts attired in training versions of their Shuttle launch and entry suits participated in many training sessions in the Fixed-Base Shuttle Mission Simulator in the Jake Garn Simulation and Training Facility (Building 5). The simulators consisted of high-fidelity mock-ups of the Shuttle's flight deck, as well as a low-fidelity mock-up of the middeck. Simulation software modeled all systems, including many preprogrammed malfunctions, response to cockpit controls, and interactions between systems.

For more than 30 years, astronauts have spent hours training in the T-38 aircraft. Some crew members considered flying in the sleek white jets one of the most important aspects of their training, as the planes provide a teaching environment where real-life situations require astronauts to make critical decisions in an uncontrolled environment, unlike the safety of a simulator. NASA's small fleet is housed at nearby Ellington Field.

Nearly one million people crowded Florida beaches and highways to catch a glimpse of history on July 8, 2011, the last Space Shuttle launch. White Flight Control Room personnel in Houston (above) monitored STS-135 as the mission delivered supplies to the ISS. When *Atlantis* undocked from the station, Shuttle commander Christopher Ferguson reflected, "When a generation accomplishes a great thing, it's got a right to stand back and for just a moment admire and take pride in its work. As the ISS now enters the era of utilization, we'll never forget the role the Space Shuttle played in its creation." Below, on July 21, 2011, the flight controllers in Houston watched *Atlantis* land in Florida, bringing the Shuttle program to a close.

An estimated 4,000 people arrived at the Johnson Space Center before dawn on July 21, 2011, to commemorate the "final stop" of the Space Shuttle *Atlantis*. Gathered on the front lawn of Building 1, employees, families, and area residents watched the last Shuttle landing on a jumbo-sized television screen. Later, Center Director Mike Coats (standing) ceremoniously lowered the *Atlantis* flag from the site's flagpole for the last time.

After the successful completion of their mission, the four crewmembers of STS-135 returned home to Houston. Hundreds of people braved the hot and humid summer day to greet the astronauts at Ellington Field's Hangar 990. The hangar areas were used as a public venue to welcome back crews throughout the years of human spaceflight.

For 23 years, fresh roses sat in the midst of the flight control room during every Shuttle mission. The Shelton family quietly started the tradition in June 1988 to show support for the Return to Flight mission after the *Challenger* accident. The bouquet included a rose for each astronaut on the crew, plus a single white rose to remember the astronauts whose lives were lost in the quest for exploration.

The Mission Control Center in Building 30 was renamed in 2011 in honor of Christopher C. Kraft, NASA's first flight director. Kraft began working as an engineer for NASA's predecessor, the National Advisory Committee for Aeronautics, in 1945. His leadership in Mission Control was critical to the success of the Mercury, Gemini, and Apollo missions. Kraft then served as the center director from 1972 until his 1982 retirement.

Flight controllers monitor the consoles in Flight Control Room (FCR) 1 in the Mission Control Center 24 hours a day, seven days a week. The Expedition missions to the International Space Station require constant support from the ground in order to maintain humanity's outpost in space. In the 1960s and 1970s, flight controllers in Mission Operations Control Room 1, as it was then known, monitored the Saturn 1B vehicle launches for Apollo 7, Skylab, and the Apollo-Soyuz Test Project. During the Shuttle program, the room was renamed FCR-1, and it became the primary control room for most of the unclassified Shuttle flights through STS-76 in 1996. The consoles and multi-tiered configuration were then replaced with new workstations and equipment as the room was converted to the Life Sciences Center. Due to the growth of the International Space Station program and the cooperation required among international control centers, FCR-1 underwent another major renovation and equipment update in 2006 to become the new home of the ISS flight controllers.

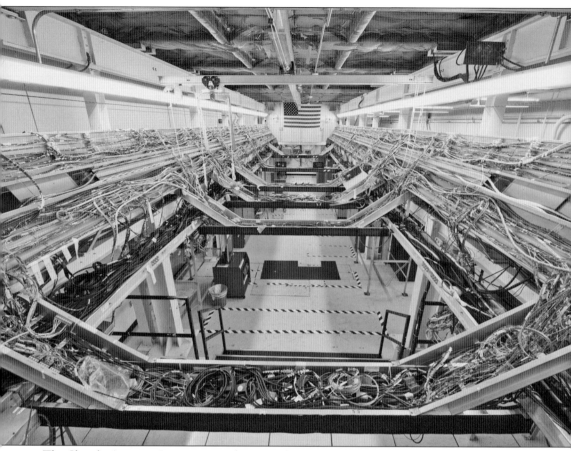

The Shuttle Avionics Integration Laboratory (SAIL) in Building 16 was used to verify all the avionics hardware and software on the Shuttle before it was sent up to space. The SAIL, also known as orbital vehicle (OV) 95, replicated the Orbiter's cockpit, middeck, and payload bay so astronauts and simulation teams could practice flying missions in both normal and emergency conditions while ensuring there were no bugs in the Shuttle's computer systems. SAIL personnel often worked long hours, sometimes staying overnight to support the 24-hour operations. After 30 years of faithful service to the Shuttle program, in July 2011, the SAIL completed its final runs of flight avionics systems. The experience gained in the three decades of operations helped lay the foundation for the Crew Exploration Vehicle Avionics Integration Laboratory, designed to support future missions to Mars.

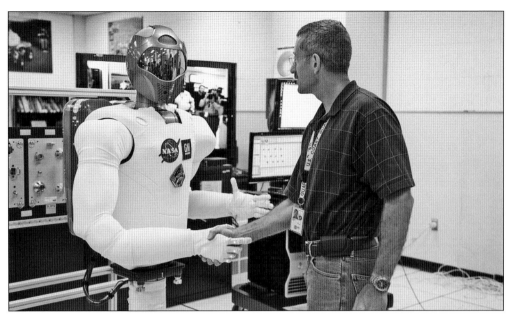

The Dexterous Robotics Laboratory in 1997 began developing a humanoid machine to work alongside astronauts. Ten years later, NASA teamed with General Motors to produce Robonaut2, or R2, which was delivered to the International Space Station in February 2011 by the STS-133 crew. NASA's experience with R2 on the station will help determine its capabilities for use on future missions to deep space.

NASA's Desert RATS (Research and Technology Studies) evaluate technology, human-robotic systems, and spacewalk equipment in Arizona's desert to mimic conditions that future astronauts may encounter on other surfaces in space. The team also uses tools and simulators in Building 9 to conduct tests in simulated microgravity, such as a mock Multi-Mission Space Exploration Vehicle (above) that moves across an air-bearing floor while set in its "flying" configuration.

Astronauts participate in physical training sessions at the Center for Human Spaceflight Performance and Research, Building 37. Many use the Advanced Resistive Exercise Device, designed as a long-term solution to preserve muscle strength during extended living in microgravity environments. Software guides crewmembers through a customized regimen while simultaneously allowing personnel on the ground to monitor.

Trainers in Building 585 worked with International Space Station crewmembers (shown above) on the use of photographic equipment. Since the earliest days of human spaceflight, astronauts have learned to capture scientific observations of geological, oceanographic, environmental, and meteorological phenomena through the camera lens. Some critical environmental sites have been regularly photographed since the Gemini and Skylab missions in order to provide long-term data.

Four distinct laboratories in Building 17 comprise the Space Food Systems Laboratory. In this multipurpose facility, space foods are designed, developed, and evaluated by a team of food scientists, registered dietitians, packaging engineers, food systems engineers, and technicians. Safety standards developed by the JSC lab through the years for food production are used by industry by the US Food and Drug Administration and the Agriculture Department.

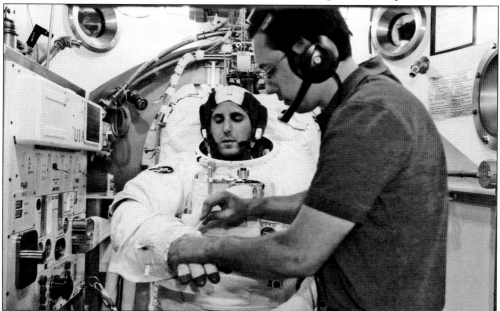

The Extravehicular Mobility Unit (EMU) serves as a personal life support and communications system for the astronaut while working outside the spacecraft. Prior to flight, astronauts undergo "fit checks" in the Space Station Airlock Test Article in Building 7 to ensure that the EMU's various components fit together securely. Engineers continue to develop spacesuit technology in preparation for long-duration missions to unexplored areas of space.

The Robotics Group demonstrated the Small Pressurized Rover, a mobile habitat module, on the lawn of Building 1 in 2009. The rover allows crews of two to four astronauts to explore the surface of new planets while away from their primary base for three days at a time. Astronauts can work in a comfortable shirtsleeve environment, using the cockpit's wide field of view to observe an extraterrestrial landscape.

The Space Exploration Technologies Corporation, or SpaceX, displayed its reusable Dragon spacecraft (left) at the 2011 NASA Innovation Day. The annual event showcases cutting-edge research from departments around the center. JSC manages the Commercial Crew and Cargo Program Office for NASA, which coordinates efforts with companies in private industry to develop the next generation of reliable and economic space transportation technologies.

Strategic partnerships between the Johnson Space Center and emerging technology companies in the Houston area help support NASA's long-term goals of increasing collaboration between the public and private sectors. Discussing plans for future cooperation are, from left to right, Center Director Mike Coats, Houston Technology Center Chief Executive Officer Walter Ulrich, and Mayor Annise Parker of Houston.

The Innovation Design Center in Building 348 opened in 2011 to serve as a creative space where employees at the Johnson Space Center can develop ideas. The facility consists of two sections, a collaboration area and a fabrication shop, to encourage innovation and professional development. These types of initiatives stimulate the visionary thinking that marks JSC as a leader in science and technology.

NASA's Multi-Purpose Crew Vehicle, referred to as the Orion crew module, undergoes testing in the 6.2-million-gallon Neutral Buoyancy Laboratory. The spacecraft is designed to accomplish NASA's goal of sending astronauts to an asteroid by 2025 and onward to Mars by the mid-2030s. Orion measures just 15 feet across and 15 feet high and is currently configured to carry four astronauts into space for up to 21 days. Unlike the Space Shuttle, which landed on the Earth like a plane, Orion will splash down in the ocean. Water landing reduces the amount of hardware needed on the vehicle, resulting in a lighter spacecraft that can journey deeper into the solar system than ever before. This also provides the crew with more space to live and work onboard. Orion will be NASA's first new vehicle for human spaceflight since the Space Shuttle *Endeavour* joined the fleet in 1991.

NASA engineers look to the future as they design outposts that can be used for research, testing, storage, and living in space. The inflatable habitat (above) expands into an airlock-sealed capsule that can be efficiently carried to remote regions of the solar system. An electric rover (pictured outside Building 220) will dock with the habitat to provide direct and safe access to the facility.

Engineers at the Johnson Space Center developed Project Morpheus in collaboration with private industry to serve as a test bed for new and cheaper propulsion, guidance, and control technologies. The Morpheus vertical lander is designed to haul about 1,100 pounds of research equipment into space and may be the next vehicle to land on the Moon, Mars, or an asteroid in advance of a human crew.

A human presence on the International Space Station has been ongoing since November 2, 2000. This photograph shows the orbiting outpost and zero-gravity research facility in its final configuration. During construction from 1998 to 2011, astronauts conducted 161 spacewalks and spent a combined 1,021 hours assembling the station. The complex spans the area of an American football field and weighs 861,804 pounds, although it is weightless in the environment of space. The station has more livable area than a conventional five-bedroom house and includes

two bathrooms, a gymnasium, and a 360-degree window. For the next decade, the international crews of the ISS will conduct research in medicine, materials, and science to further humankind's journey into the cosmos and help improve life on Earth. The Johnson Space Center in Houston is responsible for managing ISS operations, coordinating with five space agencies representing 15 nations across the globe.

Discover Thousands of Local History Books
Featuring Millions of Vintage Images

Arcadia Publishing, the leading local history publisher in the United States, is committed to making history accessible and meaningful through publishing books that celebrate and preserve the heritage of America's people and places.

Find more books like this at
www.arcadiapublishing.com

Search for your hometown history, your old stomping grounds, and even your favorite sports team.

Consistent with our mission to preserve history on a local level, this book was printed in South Carolina on American-made paper and manufactured entirely in the United States. Products carrying the accredited Forest Stewardship Council (FSC) label are printed on 100 percent FSC-certified paper.